The Green Phoenix

The Green Phoenix

A History of Genetically Modified Plants

PAUL F. LURQUIN

Columbia University Press

New York

Columbia University Press
Publishers Since 1893
New York   Chichester, West Sussex

Copyright © 2001 Columbia University Press
All rights reserved

Library of Congress Cataloging-in-Publication Data
Lurquin, Paul F.
    The green phoenix : a history of genetically modified plants / Paul F. Lurquin.
        p. cm.
    Includes bibliographical references (p.).
    ISBN 0-231-12262-4 (cloth : alk. paper)—ISBN 0-231-12263-2 (pbk. : alk. paper)
        1. Transgenic plants—History. 2. Plant genetic engineering—History. I. Title
SB123.57.L87 2001
631.5'233'09—dc21   2001017261

Casebound editions of Columbia University Press books are printed on permanent and
durable acid-free paper.
Printed in the United States of America
c 10 9 8 7 6 5 4 3 2 1
p 10 9 8 7 6 5 4 3 2 1

*To Linda Stone and Sharobi Watson, who taught
me that life is profound and fun, and to the
memory of Georges Gysels, who taught me
that so is history*

Contents

Preface

In his excellent little book *The lac Operon: A Short History of a Genetic Paradigm*, Benno Müller-Hill (1996) complains that "molecular biology has no history for the young scientist." He goes on to say, "Old errors in interpretation are not mentioned. Who cares? There is only one view, and this is the correct, modern view." How right he is! For example, I invariably notice at the beginning of my course in general genetics that undergraduate students respond as if the science of genetics had created itself by an act of spontaneous generation, with all the right answers to all the right questions, right from the start. These students should of course not be blamed; few professors and teachers have told them differently, in any area of science.

Similarly, textbook science is a very poor representation of the scientific process as it is experienced in the laboratory. In these books, faceless humans generate hypotheses easily and invariably verify them experimentally. In the field of genetics, students memorize the names of Mendel, Watson, and Crick but remain unaware of the thought processes leading to their discoveries. False starts and wrong interpretations usually are not even mentioned and are certainly not discussed, and the chronological development of ideas is largely ignored. In other words, students are rarely, if ever, exposed to the inner workings of the scientific process and typically receive a highly sanitized version of the act of scientific discovery.

Even graduate students, eager to learn cutting-edge principles and technologies, seem insufficiently aware of the sometimes convoluted process that led to the knowledge they are so keen on acquiring. In this case, however, interest in the scientific process itself is much higher. After all, these young people are tomorrow's scientists. Yet they are rarely exposed formally to the ups and downs over time of their field of choice. As for the general public, how can it hope to understand where scientific verities (and counterverities) are coming from in this age of sensationalizing mass media and sound bites. Everything is glossed over, save the most prurient.

The field of plant genetic engineering (or of genetically modified plants, as it is now incorrectly but universally known) fits the preceding description: Who remembers its birth and development? Who knows the science behind today's applications? But then, who does not know about the existence of genetically modified crops? Thus, the aim of this book is to provide a narrative of the origins of plant biotechnology, including the scientific controversies created early in this endeavor. These rocky beginnings led to great achievements, and a new wave of controversy ensued, this time surrounding the use and potential ecological impacts of genetically modified plants. Thus, what seemed in the beginning to be just another innocent (and perhaps ultimately useless) pursuit of knowledge turned into a multibillion-dollar industry, contributed to riots against the World Trade Organization summit in Seattle in December 1999, caused debates involving common people and royalty, and inflamed passions in many countries. And this story continues to unfold.

In this volume, I strive to paint an objective, complete picture of the development of ideas that have driven the scientific process underlying all present applications of gene technology in the plant kingdom. I hope this work will show again that, to paraphrase Louis Pasteur, science progresses by leaps and bounds but sometimes takes a few steps back before resuming its course forward. The developing field of plant genetic engineering was no exception to this statement.

This book is intended for advanced undergraduates, graduate students, young professionals, and others with an interest in the scientific intricacies and history of plant genetic engineering. It requires knowledge and understanding of elementary classical and molecular genetics. Technical appendices help to jolt faltering memories or to describe methods rarely in use today but heavily relied on in yesteryear.

I thank my colleagues, Professors Milton Gordon (University of Washington, Seattle), László Márton (University of South Carolina, Columbia), and Charlotte Omoto, Linda Stone, and Diter von Wettstein (all from Washington State University, Pullman) for their encouragement and for reading parts or all of the manuscript. My deepest gratitude goes to Dr. Mary-Dell Chilton (Novartis) for her detailed critique of my manuscript and her encyclopedic knowledge of the field. I have often heard young scientists and graduate students ask the sexist question, "Who is the 'father' of plant biotechnology?" My answer is that there really is no single "father," but I can without hesitation propose the name of the "mother": that of Mary-Dell

Chilton. Mary-Dell and I certainly agree on the science that led to plant biotechnology applications, much of which she developed herself. However, she and I part ways somewhat, without any animosity whatsoever, when we consider the role of biotechnology corporations in the world. Academia and industry do not always meld perfectly, and perhaps they should not. I think she understands.

I am also grateful to Denise Palmen (Technical Services, Washington State University) and to her skill with computers for salvaging old photos and diagrams that seemed totally hopeless. Finally, I thank Holly Hodder, publisher at Columbia University Press, for encouragement and support during all phases of the publishing process, and Marjorie Wexler, my copy-editor. All errors, omissions, and biases in this book remain mine.

Introduction

C

Since the discovery of the molecular nature of genetic material in 1944, the confirmation in 1952 that it is indeed DNA, and the discovery of the double helix in 1953, geneticists have been fascinated with the idea of using pure DNA to modify the genetic characteristics of organisms. Few, however, believed that this goal could ever be achieved outside the realm of a few bacterial genera, and relegated the concept of DNA-mediated genetic transformation to science fiction. Not until the late 1970s were eukaryotic cells, fungal and mammalian, shown to incorporate pure DNA fed to them, and to express the genes present in it. Twenty years later, the phenomenon of transgenesis, as DNA transformation is now known, seems routine to many. What is controversial today is the *use* of genetically engineered plants and animals, not whether they can be produced. Yet, not so long ago, starting in the late 1960s, this is exactly where the question was: Would it one day be possible to generate plants harboring totally foreign genes and expressing them?

Transgenic plants have been a reality since 1983. To the casual and innocent observer, it may seem that this branch of science developed just like any other, slowly but smoothly at first, and then exponentially. The former is definitely not true. The early attempts at transgenesis in plants were chaotic, too fast, and yielded irreproducible results for a period of at least 10 years. In the end, the scientific process did win, however. Current textbooks do not and perhaps should not get into any of the raging controversies that surrounded early claims of plant genetic transformation. However, simply ignoring this period of history does not do justice to the scientific process and does not show to young researchers that interpretations gone terribly wrong still have a heuristic value for the stimulating effects they have on scientific thought. Some of the debates that took place in the mid seventies were Homeric, frustrating, and exhilarating, all at the same time. This was a period of paradigm shifting in the then sedate field of plant biology.

This book covers the long gestation period preceding, and the time immediately following, the first solid evidence that indeed, like many organisms in all other kingdoms, plants can be genetically altered through horizontal gene transfer. Chapters 1 and 2 cover the controversial beginnings of the field, and chapters 3 and 4 describe the scientific efforts that led to the successful approaches still in use today. Chapter 5 goes into some unresolved problems associated with the genetic engineering of plants, as well as a brief analysis of the societal impact that these plants have created and will certainly continue to generate.

Undoubtedly, the first pioneer who took seriously the possibility that plants could be genetically modified through uptake of purified DNA was Lucien Ledoux of Mol, Belgium. The two most important contributions he made to the field were published in the prestigious journals *Nature* and *Journal of Molecular Biology*. They were the claim that foreign DNA could become integrated and replicated in barley (Ledoux and Huart 1968, 1969) and the claim that *Arabidopsis* mutants could be genetically corrected by prokaryotic DNAs carrying the relevant genes (Ledoux et al. 1974). One could then wonder why it took so long, over 10 years, for plant genetic transformation to become generally accepted. After all, no one today doubts that foreign DNA can become integrated within plant genomes and replicated along with the host DNA, and indeed, transgenic plants do express completely foreign genes. The answer is deceptively simple: These results could not be duplicated by others, even in Ledoux's own laboratory, although some laboratories in Australia (Doy et al. 1973) and Germany (Hemleben et al. 1975) had made similar claims in other plant systems. Almost equally controversial results concerning the effects of exogenous DNA on flower color in *Petunia* had been published during the same period by Dieter Hess of Germany. Taken together, these early positive results seemed consistent and mutually reinforcing, although they raised questions regarding the techniques used and data interpretation. By 1976, however, it had become clear that claims of foreign DNA integration and expression in plant cells were on very shaky grounds; negative evidence obtained by more sophisticated methods was becoming overwhelming. Some even predicted the demise and death of the whole field.

By pure chance, 1976 was also the year when, for the first time, it was conclusively demonstrated that *Agrobacterium tumefaciens*, the bacterial agent that causes crown gall disease, was able to transfer spontaneously, it seemed, a portion of its own DNA to recipient plant cells (Chilton et al.

1977). Thus, what appeared to present intractable problems in the lab had been done all along by Nature itself! Also, what Nature was doing could possibly be imitated and even perhaps improved upon by humans. And indeed, transgenic plant production through *A. tumefaciens*–mediated gene transfer has become the most widely used technology in this field. Further, the discovery of DNA transfer from *Agrobacterium* to plant cells is what allowed the field of direct (naked DNA–mediated, and independent from *Agrobacterium*) gene transfer a quick rebirth and eventual success.

All events and scientific results described in this book are strictly documentable. However, I could not possibly mention all the articles published over a period of roughly 20 years, starting in 1968; the list of references might then be longer than the text. Rather, I will focus on the contributions of groups and cite their seminal papers. In particular, but certainly not exclusively, I will focus on the work of the Mol (Belgium), Hohenheim (Germany), Seattle (United States), St. Louis (United States), and Ghent and Ghent/Cologne (Belgium/Germany) groups; the Leiden (The Netherlands) and Monsanto (United States) groups; and finally the Nottingham (United Kingdom) group. The Basel (Switzerland) team should receive special mention because this is where the first incontrovertible evidence for stable DNA–mediated plant transformation was obtained (Paszkowski et al. 1984).

It should be noted that Chapter 5 is not strictly limited to the science of plant transgenesis. In it, I present some personal views and ideas on the role of multinational biotechnology companies in the world, and the potential impact of plant transgenesis applications on human welfare. These thoughts should not be misconstrued as rejecting all of biotechnology and its industrial applications. Rather, they represent my own ambivalence about this powerful new technology. On the other hand, I do not hesitate to criticize and reject the actions of vandals that some have called ecoterrorists. Clearly, some of my views and ideas will not be to everyone's liking.

This book will show, among other things, that scientists can sometimes be fooled by their own and others' results, and that it can take a long time for misinterpretations to be corrected. Also, it will show that bold new claims with high-stake implications, such as free DNA–mediated genetic transformation as published in the early days, are much more susceptible to very close and more general scrutiny than run-of-the-mill reports. This is reminiscent of an article by Gunther Stent (1972), in which he discusses the notions of prematurity and uniqueness in the sciences and uniqueness in the arts. For Stent, "A discovery is premature if its implications cannot be connected by a series of simple logical steps to canonical, or generally accepted, knowledge." That much can certainly be said of early claims about plant transformation, and this is why these claims triggered an avalanche of criticism. However, these criticisms eventually led to great success *because* they allowed a deep rethinking of the transformation process. In that sense, plant genetic transformation was not at all characterized by uniqueness; rather, the highly interactive and competitive nature of modern science created opportunities for multiple and incisive analyses of these bold new claims. Meanwhile, the elucidation of *Agrobacterium*-mediated gene transfer, itself involving dozens of researchers in several countries, cross-fertilized a struggling field and allowed it to fructify. Thus, it cannot be said in the case of plant transgenesis that a single unique creative act was at the origin of plant transformation. On the contrary, the scientific pendulum swung wildly for a significant period of time and, like a phoenix, transgenesis in plants had to die before it could be reborn.

CHAPTER I

Where It All Began

𝓬

The community of science is very much unlike any other. Academic scientists do research on strange and often apparently useless topics, publish their results in journals that nobody else reads, and go to meetings that interest only themselves. Harmony seems to reign. At least, this is what a superficial examination reveals. The history of plant genetic engineering is actually characterized by deep controversies as well as great achievements. Truth in the area of plant transformation did not prevail right from the start, and it took years for some claims to be universally accepted or rejected. The beginning of this book thus describes the very first attempts at plant transgenesis, which took place in a remote location in Belgium, the town of Mol. This chapter will show that science can sometimes be a chaotic phenomenon and that scientific truth is not necessarily like Botticelli's Venus: It does not emerge spontaneously from the surf of experimental work.

Genesis

Because the beginnings of plant transformation research can be traced to one single laboratory, I believe that a short description of the place and circumstances is warranted. Mol is a strange little town. Were it not for the presence of three huge nuclear reactors in a nearby fenced-in conifer wood, it would resemble any other small community in eastern Antwerp province in the vicinity of the Dutch border. Towns in this area had always been farming communities condemned to cultivate arid, sandy soils or, in the case of nearby Geel since medieval times, to specialize in the reinsertion into society of mentally disturbed patients. The proximity of Geel was, of course, an endless source of jokes. The building of the Nuclear Study Center, however, changed the atmosphere of the place in more ways than one. First came the massive concentration of fissile materials. As these were the late 1950s, the inhabitants did not seem too much troubled. After all, nuclear energy was good back then. Next came highly educated people whose function it was to operate the reactors (one for research, one for electricity production, and the third for research and radioisotope synthesis) and, as in U.S. nuclear centers (of which the Belgian one was a copy), scientists in biological fields came to work in a radiobiology department. The original mission of this department was to do research on the peaceful uses of nuclear energy and, in particular, on radioprotectants, chemicals destined to shield people from the harmful effects of ionizing radiations during medical treatment or, heaven forbid, to protect them for intentional or unintentional fallout. Curiously, this department devoted to radiobiology also housed the laboratory of Lucien Ledoux, who had embarked on a tremendously bold project, the genetic transformation (that is, DNA-based genetic engineering or transgenesis) of living cells and organisms at large, from prokaryotes to plants to mice and rats. Because this book is about transgenesis in plants, his research and that of his coworkers on bacteria and mammalian cells will not be discussed here, except to note that, contrary to his work with plants, his results with prokaryotes and mammalian cells have had very little impact on the scientific community.

Transforming plants was a far from trivial endeavor: The reader should remember that at that time, only gram-positive *Bacillus subtilis*, *Hemophilus influenzae*, and *Streptococcus pneumoniae* could be routinely transformed with DNA. It turns out that these three prokaryotes take up DNA spontaneously, through a complicated process that allows the entrance of single-stranded DNA only. What is more, there was no evidence (and there still is none) that these organisms were transformable by nonrecombinant, completely heterologous DNA (transformation *within* the genus *Bacillus*, however, is possible, depending on the percentage of homology between the recipient and the host). Compounding these technical difficulties was the extreme reluctance of the scientific world to accept what is today called horizontal or lateral gene transfer. It was simply inconceivable that unrelated living cells could exchange genes, as this would violate the idea of slow, mutation-driven evolution and the existence of strict sexual barriers. One of the major arguments used to buttress this view was that humans consumed vast amounts of plant materials, and yet nobody had so far become a photosynthetic organism after eating lettuce. Similarly, one did not turn prokaryotic after dipping a finger into an *Escherichia coli* DNA solution. In many ways, most scientists found these ideas of heterologous gene transfer completely silly and perhaps even anarchical.

Thus, there existed at the time no technique promoting DNA uptake by any type of cells that could not become spontaneously competent, including *E. coli*, the workhorse of bacterial genetics. This also meant that there existed no simple model transformation system on which one could base further experimental work. In addition, in the case of plants, genetic markers and in particular auxotrophic, nutritional mutations were conspicuously almost totally lacking, making experiments on genetic complementation (correction of mutations to make the effects of foreign DNA detectable) a real challenge. Therefore, the emphasis had to be put on physical techniques allowing the researcher to trace the fate of donor DNA, such as its degradation (or not) and its intracellular distribution, in the recipient cells. Such techniques could, of course, neither prove nor disprove any biological effect of foreign DNA, but at least they could potentially give information regarding the location and physical state of this DNA in the recipient plants or other organisms.

This could be done by feeding radioactive bacterial DNA to these organisms and following its presence or absence in various organs and cells. Why was bacterial DNA used? First, DNA of high specific radioactivity had to be used, or the recipient cells would have had to be drenched with it. Only bacterial mutants deficient in the synthesis of nucleic acid precursors could achieve that goal. Because all cells naturally contain DNA, it was necessary to differentiate donor from endogenous DNA in the treated cells or organisms, and this, it seemed, could easily be done if the donor DNA had a different cesium chloride (CsCl) buoyant density (d) (i.e., a different G + C composition) from the host DNA (see appendix 1). Only bacterial DNAs afforded this difference, from *Micrococcus lysodeikticus* (also called *M. luteus*) ($d = 1.731$) to *Clostridium perfringens* ($d = 1.691$) and just about everything in between, considering also that bacteria could be grown in deuterated (heavy) water, their DNA becoming denser accordingly. Eukaryotic nuclear DNA has a density of around 1.700, give or take a few 0.001 density units, meaning that density differences would be easily detectable. Hence, DNA-treated plant cells and whole plants were ground up and their DNA was extracted and analyzed by CsCl buoyant density gradient centrifugation. The presence in the gradients of DNA peaks distinct from that of the endogenous DNA would give information regarding the uptake and physical state of the foreign DNA. For example, the detection of a narrow peak banding at the density of donor bacterial DNA would indicate the presence of high-molecular-mass donor DNA, whereas a broad peak at that density would be indicative of some donor DNA breakdown after contact with cells. Radioactivity banding at the level of host DNA could mean that the donor DNA had been broken down into small subunits subsequently reused for host DNA synthesis. DNA peaks (if any) at other locations in the gradients clearly deserved other explanations. It should be remembered that DNA buoyant density in CsCl provides no information at all regarding the *sequence* of the analyzed molecules.

What, then, were the preliminary results obtained by the Mol team? First reports, including one published in *Nature* in 1966 (Stroun et al. 1966), indicated that exogenous radioactive bacterial DNA could be translocated to various plant organs in tomato and barley, presumably via the vascular system. In and of itself, this observation was not revolutionary, but it suggested that a macromolecule could transit through intercellular channels and perhaps penetrate cells. Here I use the term *suggest* because no efforts were made to get rid of, by deoxyribonuclease (DNase) treatment, DNA loosely bound to the outside of cells. No evidence was given that the donor DNA was actually intracellular other than the fact that autoradiography experiments showed the presence of label in nuclei. Because CsCl gradients showed that the majority of the radioactivity cosedimented with plant DNA, nuclear labeling was to be expected because of the breakdown and reutilization of the donor DNA for host DNA synthesis. Since *in situ* DNA-DNA hybridization did not exist in 1967, the presence of polymerized donor DNA anywhere in the system (other than simply associated with cell walls, for example) could not be ascertained.

These experiments were performed in collaboration with two Swiss scientists, Philippe Anker and Maurice Stroun, and another member of the Mol staff, Pol Charles. Further work by Anker and Stroun (1968), also published in *Nature* two years later, purported to show that tomato shoots incubated in the presence of radioactive *A. tumefaciens* DNA and subsequently analyzed for DNA banding in CsCl gradients, not only showed the presence of donor DNA (as in the previous *Nature* paper), but also indicated that the tomato DNA, when rebanded and sheared by sonication, released molecules with the buoyant density of *A. tumefaciens* DNA. The conclusion was that the label found at the level of tomato DNA was not simply a result of donor DNA degradation and reutilization for endogenous DNA synthesis, but furthermore that part of this label was the result of the *physical integration* of long stretches of *A. tumefaciens* DNA within tomato DNA (fig. 1.1 and appendix 1). Indeed, the rationale was that shearing by sonication reduced the size of the DNA molecules, and then pieces of bacterial DNA originally covalently linked to plant DNA would be released and would band at the density of *A. tumefaciens* DNA.

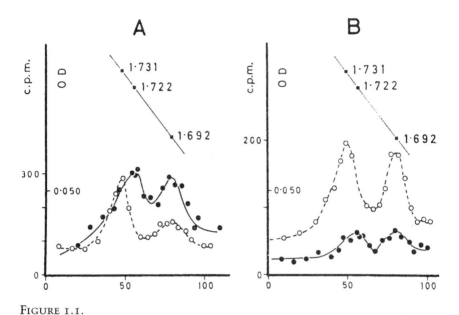

FIGURE I.I.
First experimental results used to infer integration of bacterial DNA into tomato DNA. A: CsCl ultracentrifugation analysis of DNA isolated from tomato shoots incubated with radiolabeled DNA from *A. tumefaciens*. B: The peak banding at the level of tomato DNA ($d = 1.692$) was harvested, sonicated, and rebanded in a CsCl gradient. This resulted in band splitting and the appearance of a peak at the level of the donor DNA. *Open circles*: UV absorbance, which includes that of a density marker (*M. lysodeikticus* DNA, $d = 1.731$). *Closed circles*: Radioactivity. The ordinate represents fraction number. (From Anker, P., and M. Stroun. 1968. Bacterial nature of radioactive DNA found in tomato plants incubated in the presence of bacterial DNA-[3]H. *Nature* 219:932–933, figure 3. Copyright *Nature*. Reprinted by permission of the publisher.)

This must have been perceived at the time as the ultimate blockbuster. For the first time ever, a potential recombination event between prokaryotic DNA and plant DNA had seemingly been demonstrated: Naked DNA could be taken up by plant cells, where it became physically integrated within the recipient genome. Many would have to agree that today a journal like *Nature* would not so easily publish work of such enormous impact. No evidence was provided that the *A. tumefaciens* DNA peak was not caused by cross-contamination, as no rebanding of the host DNA free from donor DNA was shown or even mentioned in the text. Further, even though a DNA-DNA hybridization technique had been published two years earlier (Denhardt 1966), no effort was made (or at least published) to apply it to these extraordinary molecules. Moreover, the tomato shoots used as recipients were not checked for the presence of microscopic contaminants whose own DNA might have influenced the results.

However, there was more to come. Persistence and incorporation of foreign donor DNA in plant cells are one thing, replication of donor DNA would be quite another, as this would open the door to the possibility of stable genetic expression and genome modification of plants.

We will now enter the realm of plant genetic engineering 16 years before it was actually demonstrated in a way accepted by the scientific community. This time, Anker et al. (1967) fed *nonradioactive* bacterial DNAs to tomato shoots for 48 hours and followed this treatment by labeling with ^{3}H-thymidine. DNA was then extracted and analyzed by CsCl density gradient centrifugation. To what must have been everyone's surprise, the radioactivity, far from being concentrated solely in the tomato DNA peak ($d = 1.692$), was present exclusively in a peak banding at an *intermediate* position between the donor bacterial DNA and tomato DNA. For example, when *A. tumefaciens* DNA (reported as $d = 1.722$, whereas the true density is 1.718) was used as donor, almost all the ^{3}H radioactivity banded at $d = 1.707$. Further, shearing of this intermediate peak by ultrasonication yielded *two* peaks in CsCl gradients, one with the density of the donor DNA and one banding at the position of tomato DNA. The authors' conclusion was that "DNA molecules of intermediary density are formed as a consequence of the attachment of the foreign double-stranded DNA to the native endogenous one" (Anker et al. 1967). Thus, as in the previous article, integration of donor bacterial DNA was claimed, but this time, this cointegrate structure was also able to replicate. No controls with ^{3}H-thymidine labeling in the absence of pretreatment with bacterial DNA were done, but treating the shoots with *Clostridium perfringens* DNA ($d = 1.691$) yielded no intermediate peak, as expected. Further, the density of the intermediate peak was proportional to that of the donor DNA—that is, the heavier the donor DNA, the heavier the intermediate band (fig. 1.2). As before, the tomato shoots were not shown to be axenic.

It is now time to pause and reflect on these extraordinary results. To summarize, purified heteropycnic (having a density different from that of the host) bacterial DNA molecules can be translocated to various organs of a plant shoot, presumably penetrate the cells, penetrate the nuclei of these cells, and become joined with endogenous DNA, and these structures can also replicate. Except perhaps for the translocation part, this statement reads like a description of modern DNA-mediated gene transfer. However, did integration and replication of the foreign DNA really happen? A close study

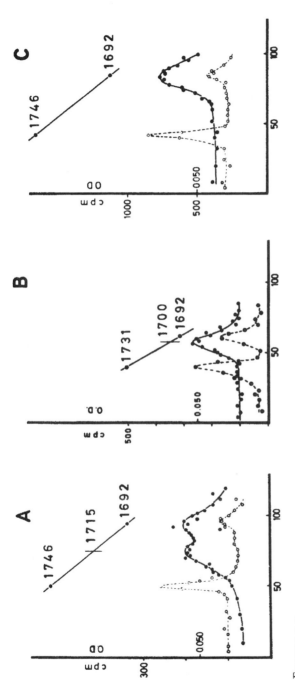

FIGURE I.2.
Experiment in which three bacterial DNAs with different buoyant densities affect tomato DNA differently after incubation with shoots. In this case, shoots were incubated with unlabeled bacterial DNAs and then with radioactive thymidine, to study the replication of the foreign DNA molecules. A: Incubation with *M. lysodeikticus* DNA ($d = 1.731$). B: Incubation with *E. coli* DNA ($d = 1.710$). C: Incubation with *C. perfringens* DNA ($d = 1.691$). *Open circles*: UV absorbance, including that of denatured *M. lysodeikticus* DNA used as density marker ($d = 1.746$). *Closed circles*: Radioactivity. The ordinate represent fraction number. (From Stroun, M., P. Anker, and L. Ledoux. 1967. DNA replication in *Solanum lycopersicum esc.* after absorption of bacterial DNA. *Curr. Mod. Biol.* 1:231–234, figure 2. Copyright Elsevier Science Ireland, Ltd. Reprinted by permission of the publisher.)

of the articles by Anker and Stroun (1968) and Stroun et al. (1967) shows some absence of consistency. Indeed, the results in Anker and Stroun (1968) do not reveal the existence of a DNA band with a density intermediate between that of donor and that of tomato DNA. Why not? Then, in Stroun et al. (1967), the rate of replication of these intermediate-density DNA molecules is much higher than the rate of replication of tomato DNA. Figure 1.2 shows this very clearly: There is some radioactivity located at the position of the tomato DNA in the gradient, but the specific radioactivity of the intermediate band is much higher (minimum UV absorbance at the maximum of radioactivity). What is happening here? Could it be that DNA molecules in the intermediate-density peak replicate preferentially? Is the synthesis of these molecules taking place inside tomato nuclei, or do they correspond to replicating DNA present in contaminating microorganisms? Figure 1.2 would seem to rule out contamination, because the density of the intermediate band depends on that of the donor DNA. Of course, one could always assume the chance contamination of different shoots with different types of bacteria with different DNA densities.

It is at this point that the Swiss/Belgian collaboration ended; newcomers, both domestic and foreign, were soon to appear on the scene and revisit these interpretations. Before this happened, Ledoux and his technician Huart had published in 1969 in the *Journal of Molecular Biology* (Ledoux and Huart 1969) an extensive study on the formation of intermediate-density DNA molecules in germinating barley seeds after incubation with heteropycnic bacterial DNAs. For example, when *M. lysodeikticus* DNA ($d = 1.731$) was fed to seedlings, most of the radioactivity banded at $d = 1.712$, a value intermediate between the density of the donor DNA and that of barley DNA ($d = 1.702$). The scenario in this article was pretty much identical to that presented earlier with tomato shoots: Foreign DNA was translocated, it integrated within the host genome, and hybrid molecules replicated much faster than host DNA or suppressed its synthesis. This time, however, there was no longer any discrepancy between results obtained with ^{3}H-labeled donor DNA and results from replication experiments: Intermediate bands were observed in all cases and sonication resulted in band splitting, regenerating donor and host DNA. Furthermore, the CsCl profiles looked much more professional (fig. 1.3) and a quantitative correlation was established between the appearance of the intermediate-density band and the simultaneous decrease in host DNA specific radioactivity

A tentative explanation was offered to account for this bizarre phenomenon: that with increasing time "there exists in the population of cells having absorbed the exogenous DNA, progressively fewer individuals able

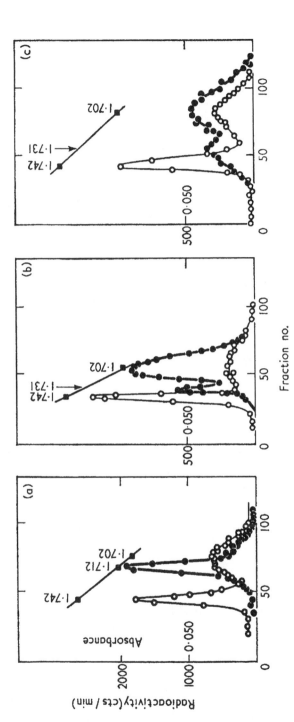

FIGURE 1.3.

Experiment aiming to show covalent integration and replication of bacterial DNA in barley. Barley seedlings were incubated with unlabelled M. lysodeikticus DNA (d = 1.731) and subsequently with radioactive thymidine. Barley DNA bands at d = 1.702. A: DNA isolated from roots was banded in a CsCl gradient. B: CsCl gradient analysis of the d = 1.712 peak from (A) after sonication for 20 sec. C: After sonication for 180 sec. Open circles: UV absorbance, including that of denatured M. lysodeikticus DNA (d = 1.742) used as density marker. Closed circles: Radioactivity. (From Ledoux, L., and R. Huart. 1969. Fate of exogenous bacterial deoxyribonucleic acids in barley seedlings. J. Mol. Biol. 43:243–262, figure 8. Copyright Academic Press. Reprinted by permission of the publisher.)

to synthesize normal DNA" (Ledoux and Huart 1969). This would then suggest that more and more cells were becoming transformed with time until, possibly, no cells would remain untransformed. Knowing what we now know about direct gene transfer frequencies, and taking into account that no selective pressure (that is, favoring the division of only those cells that had picked up the foreign DNA) could be exercised in these experiments, this explanation seems extraordinarily unlikely. Nevertheless, control seedlings labeled with ^{3}H-thymidine alone showed radioactivity banding nowhere but at the level of barley DNA, and sonication of these molecules did not produce any band splitting, just normal broadening as a result of depolymerization. Basically, the tomato story was confirmed with barley, and again, DNA-DNA filter hybridization was not used.

Unorthodox as all these claims may have been, they were published in prestigious journals and hence were widely circulated. Unorthodox results in and of themselves are of course the stuff of science and, if confirmed, have the potential to achieve dramatic progress. In 1969, however, reports of foreign DNA integration in plants had been neither confirmed nor refuted independently.

First Controversy

As already explained, claims of integration of donor DNA in plants relied on the single technique of CsCl density gradient centrifugation. However, exactly how reliable was this technique? For example, it was known that highly polymerized DNAs used as density markers in CsCl gradients had a nasty propensity to deform otherwise perfectly symmetrical gaussian peaks (the distribution of DNA molecules in a CsCl gradient follows a perfect gaussian curve at equilibrium) of the DNA molecules under analysis. In some extreme cases, a very pronounced shoulder, almost a second peak, was produced at or near the marker DNA band. This did not happen in the absence of a marker. This phenomenon can be attributed to viscosity effects literally "trapping" less polymerized, eukaryotic DNA, which takes a longer time to reach equilibrium than higher-molecular-mass bacterial DNA markers. These artifactual bands looked suspiciously like intermediate-density peaks seen in barley and tomato. To make matters worse, certain gaussian peak combinations could also result in artificial intermediate-density peaks. Thus, two peaks of different densities, when in appropriate proportions and with a given degree of polymerization (width), combined to form a single peak of intermediate density! These latter results were obtained in collaboration with my friend and colleague Max Mergeay through the use of an IBM computer as large as a good size truck, which was the norm in those days (Lurquin et al. 1972). Of course, none of that explained the mysterious effects of ultrasounds.

An important event would soon add to the questioning of the nature of intermediate-density DNA bands. International scientific meetings are, of course very often used as forums to discuss cutting-edge data. Such a meeting (sponsored by NATO) was held in Mol in August 1970, under the title "Uptake of Informative Molecules by Living Cells." Interestingly, researchers in the United States had attempted to duplicate the DNA integration/replication results with barley and were puzzled by what they saw. Thus, Yasuo Hotta, who at the time was working in the laboratory of Herbert Stern at the University of California, San Diego, came to the meeting with somewhat unsettling data. The Stern lab was well known for its study of DNA synthesis during meiosis in plants, and people there were intrigued by claims of foreign DNA integration and replication. It must be realized that at the time, DNA metabolism in plants was a poor parent; the systems were just too complicated for most. Nevertheless, the Stern lab tried to repeat the integration/replication results in barley, tomato, and lily meiotic cells and essentially failed. They did detect DNA molecules with unexpected densities, but these occurred only under extreme stress, such as desiccation or irradiation with x-rays, and these molecules were not present in nuclei. Healthy DNA–treated barley seedlings did not contain any unusual DNA molecules (Hotta and Stern 1971). Yasuo Hotta's single dissenting voice at that meeting did not go unnoticed, but it had very little immediate effect on the good standing of "intermediate" DNA molecules in plants.

Meanwhile, researchers at the University of Washington, Seattle, had independently investigated the fate of radiolabeled, heteropycnic *Pseudomonas aeruginosa* DNA in germinating pea seeds and tobacco cells in suspension culture and found no evidence at all for the presence of an intermediate-density DNA peak in their CsCl gradients (Bendich and Filner 1971). More disturbing yet was the finding that unusual CsCl gradient profiles were observed when plant tissues were known to be contaminated with bacteria (Bendich 1972).

About three years after the NATO meeting, DNA uptake studies in the photosynthesizing microeukaryote *Chlamydomonas reinhardi* were started in Mol. There was great potential to do some serious work on the biological effects of foreign DNA on this organism, which grows relatively fast, is unicellular, and can be plated like bacteria. Further, in addition to nutritional mutants, there existed a convenient CW 15 cell wall–less strain that could conceivably be more permeable to DNA than the wild type (fig. 1.4).

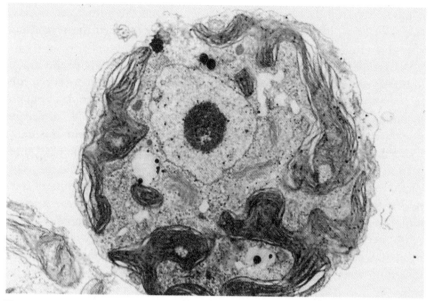

FIGURE 1.4.
Electron micrograph of the cell-wall-less *Chlamydomonas* CW 15 mutant. The cell's spherical shape is noticeable and is the result of osmotic pressure. Wild-type *Chlamydomonas* cells (with cell wall) are oblong.

It was at that time that Ram Behki, a scientist from Agriculture Canada in Ottawa, came to the Mol lab to spend a one-year sabbatical and decided to work on the *Chlamydomonas* project. A large amount of time was spent optimizing conditions for maximum donor DNA binding to the cells (a possible prerequisite for uptake), notably through the use of polycations, which were known to stimulate viral RNA uptake in higher plant protoplasts (cells devoid of a cell wall). After several months, the following straightforward conclusions were drawn: *Chlamydomonas* cells could incorporate small amounts of donor bacterial DNA, which was quickly degraded and reutilized for host DNA synthesis. No intermediate-density DNA peak was ever observed in dozens of CsCl gradients (Lurquin and Behki 1975). Further, arginine-minus mutants could not be complemented with *E. coli* DNA, and no streptomycin-resistant transformants could be recovered after incubation with DNA from a streptomycin-resistant ribosomal mutant of *Bacillus niger*. These negative genetic results were of course never published, but they suggested that biological effects of foreign DNA might not be easy to observe with the system used.

In retrospect, those genetic experiments were really naive and basically a waste of time; prokaryotic *E. coli* and *B. niger* genes could not possibly have been expressed in eukaryotic *Chlamydomonas*. However, one must remember that at that time, well before DNA sequencing, nobody knew that prokaryotic promoters were very different and probably would not function in eukaryotes. Even thinking about chloroplast genes, which possess prokaryotic-like promoters and terminators, it would have been extremely unlikely to observe an effect with *B. niger* DNA, because chloroplast transformation is now known to require homologous recombination (Svab and Maliga 1993). Finally, in the absence of cloned genes, dosage was ridiculously low. Nevertheless, it could still be hypothesized that small but biologically meaningful amounts of foreign DNA could become integrated and remain undetected (Lurquin and Behki 1975). Ironically, these optimized DNA uptake conditions were used years later by the Rochaix group in Switzerland to demonstrate for the first time genetic transformation of *Chlamydomonas* with recombinant DNA (Rochaix and van Dillewijn 1982).

By now, it had become clear that the claims of integration and replication of foreign DNA in plant cells (including *Chlamydomonas*—why not?) via formation of easily detectable cointegrate structures had possibly been exaggerated. However, absence of proof is not proof of absence, and failure to obtain a particular result could be blamed on experimental conditions. Thus, further investigations by more researchers were needed to ascertain the validity (or lack of it) of the "intermediate DNA" model.

Arabidopsis and the Comeback of Barley

It would be a mistake to believe that *Arabidopsis* as a genetic tool was discovered only in the mid 1980s. Already in the mid 1960s and early 1970s, a lot of work had been done with this small crucifer whose short life cycle and ability to grow *in vitro* made it a good laboratory organism. The name of Rédei immediately comes to mind: He had isolated and mapped several thiamin-requiring mutants that were later used by the Ledoux team to attempt genetic transformation (see next chapter). In 1971, Ledoux and associates had published an extensive study on the fate of bacterial DNA in *Arabidopsis*. As in the case of tomato and barley, intermediate-density DNA was found in plants growing from dry seeds incubated in the presence of heteropycnic radioactive DNA. As before, the intermediate-density band generated molecules with a density corresponding to that of the donor DNA upon sonication. Further, it was found, as in the case of tomato (Anker and Stroun 1968), that molecules banding at the density of *Arabidopsis* DNA, when pooled and sonicated, also showed the presence of donor DNA. The intermediate-density band was presumed to consist of DNA molecules containing long stretches of donor DNA covalently linked to long stretches of *Arabidopsis* DNA, whereas the host DNA peak contained long stretches of host DNA covalently bound to short pieces of donor DNA. This interpretation immediately poses some theoretical problems: If indeed the donor DNA is fragmented after uptake and integrated, why are discrete CsCl bands obtained at all? Assuming random breakage of the donor DNA and random integration of fragments of random lengths, one would expect that after extraction of the DNA (which also shears molecules randomly), a very broad band should be observed in CsCl gradients, corresponding to molecules consisting mostly of heavy donor DNA, all the way to molecules consisting mostly of light *Arabidopsis* DNA (or other recipient plant DNA). Such patterns were never observed. Consequently, one is led to think that fragmentation of donor DNA after uptake and integration is not random. Conversely, appearance of an intermediate-density DNA peak may have nothing to do with foreign DNA integration. For example, plant and animal satellite DNA fractions band as discrete peaks in CsCl gradients and thus mimic these intermediate-density peaks. Could it be that intermediate-density peaks are related to satellite DNA? We will see later that the answer in at least one case is yes.

An interesting property of *Arabidopsis* is its short life cycle of about 35 days. It is thus possible to examine successive generations within a reasonable period of time. The Ledoux team took advantage of this property in the following way: Dry seeds were incubated with heavy radiolabeled bacterial DNA, allowed to germinate, grow, flower, and set seed. These first generation (F1) seeds were then germinated and fruits from the F1 plants were harvested and analyzed. These fruits were found to contain radioactive DNA banding at the level of an intermediate-density peak again. This meant that the hybrid DNA molecules could be transferred to the progeny: The integrated foreign DNA was stable (Ledoux et al. 1971)! In another experiment, successive generations of seeds were "loaded" with unlabeled *M. lysodeikticus* DNA ($d = 1.731$) and in F4, plants were labeled with ^{3}H-thymidine. DNA from these plants was subjected to CsCl gradient analysis and found to contain, in addition to labeled endogenous DNA, a radioactive peak of intermediate-density ($d = 1.723$), this time detectable by its UV absorbance (this had never been seen before). In fact, the amount of DNA in that band corresponded to roughly 20 percent of the total.

A quick calculation shows that these intermediate molecules should consist of stretches of DNA containing about 75 percent *M. lysodeikticus* DNA and 25 percent *Arabidopsis* DNA. This then means that the DNA of these plants was about 15 percent bacterial. Given that the size of the *Arabidopsis* genome is about 100×10^6 base pairs, whereas a "typical" bacterial genome contains roughly 4×10^6 bp, this means that the cells of these F4 plants each contained on average the equivalent of four to five bacterial genomes, or roughly 20,000 bacterial genes per cell. It is very hard to imagine how such an organism could survive. Also, the same bizarre phenomenon that had happened in barley was observed here: The specific radioactivity of the F4 intermediate molecules was about three times that of the host DNA. Why did the intermediate molecules replicate preferentially? These questions could have been addressed by doing DNA-DNA hybridization experiments, which rely on DNA sequence rather than gross G + C content and are clearly much more specific. Of course, DNA hybridization by Southern blotting did not exist in the early 1970s and would not have been useful in the analysis of whole genomes without specific DNA probes (which did not exist either), but the nitrocellulose filter method was by now a proven technique. It was never applied, however, to the case of intermediate-density DNA molecules. What's more, a new and much more quantitative hybridization technique had been published by Britten and Kohne in 1968 (see appendix 2). This method, obsolete today because it has been replaced by Southern blots, is based on the measurement of rates of reassociation of single-stranded DNA at temperatures high enough to prevent imperfect duplex formation. Results were quantitative and graphic representation of the process took on the name of $C_0 t$ curves.

Originally, the method was used to measure genome complexity, given that the higher the sequence heterogeneity among DNA fragments, the slower the reaction rate. Thus, phage DNA of low complexity reassociated much faster than prokaryotic DNA, which in turn reassociated much more rapidly that nonrepetitive eukaryotic DNA. The technique was quickly adapted to detect homology between DNA from different sources. For example, if organism A was thought to contain DNA sequences homologous to those of organism B, the rate of renaturation of a trace amount of radiolabeled DNA from B should be accelerated in the presence of large amounts of DNA from A, and vice versa. The speeding up of the hybridization reaction could be quantified and the amount of homology between A and B calculated from the data. These came to be known as $P_0 t$ curves, where P_0 stands for probe concentration and t for time, $P_0 \times t$ being the variable plotted on the horizontal axis. Clearly, these $P_0 t$ curves would have been ideal to determine once and for all whether these intermediate-density peaks contained donor DNA integrated within plant host DNA. This is because DNA renaturation is sequence specific and not simply G + C dependent, like buoyant density in CsCl. This technique was not used either by Ledoux and his group. However, the Britten and Kohne technique was soon to see the light of day in Mol, thanks to Andris Kleinhofs, a barley geneticist, who went there on sabbatical leave from Washington State University in 1974 (fig. 1.5).

Work on bacterial DNA uptake in barley thus resumed. It quickly turned out that the technique used by Ledoux to purify barley DNA, which did not include a phenol extraction step, produced DNA of much inferior quality, as judged by the UV spectrum and molecular mass, than Kleinhofs' technique, which did include a phenol step. It was later shown by Kleinhofs et al. (1975) that sonication of impure DNA can produce multiple bands in CsCl gradients, reminiscent of band splitting seen with intermediate-density DNA. Worse, experiments aimed at verifying the presence of an intermediate-density DNA systematically failed, except when the barley seeds were contaminated with bacteria. Thus, with axenic plant material, no intermediate CsCl band was ever seen, either when the donor DNA

FIGURE 1.5.
A. Kleinhofs and the Mol group in summer 1974. From left to right, Pol Charles, Max Mergeay, Andy Kleinhofs, Paul Lurquin, Raoul Huart, and Lucien Ledoux.

was radiolabeled or when unlabeled DNA was fed to the seeds, followed by incubation with ^{3}H-thymidine to check for replication of an elusive intermediate.

How could one then account for these intermediate-density DNA peaks that had been consistently observed earlier? It turns out that, in spite of a stringent sterilization procedure, contaminated seeds were occasionally encountered and identified by incubating a small portion of the seeds on nutrient agar. These bacterial contaminants were purified, and their DNA

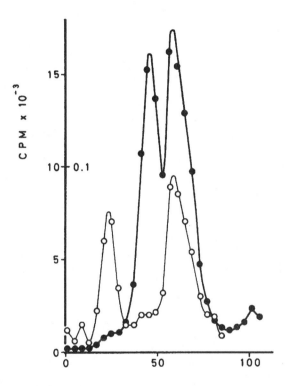

FIGURE 1.6.
Detection of "intermediate" DNA in barley contaminated with bacteria. Barley seedlings were incubated as in figure 1.3 and their DNA analyzed by CsCl gradient centrifugation. The "intermediate-density" DNA peak is on the left. *Open circles*: UV absorbance, including that of *M. lysodeikticus* DNA ($d = 1.731$) used as density marker. *Closed circles*: Radioactivity. It can be seen that here, also, the intermediate-density DNA peak displays very high specific radioactivity. The ordinate represents the fraction number. (Adapted from Kleinhofs, A. 1975. DNA-hybridization studies of the fate of bacterial DNA in plants. In L. Ledoux, ed. *Genetic Manipulation with Plant Material*, pp. 461–477. New York: Plenum Press.)

was extracted and analyzed by CsCl gradient centrifugation in the analytical ultracentrifuge. Sure enough, most of these contaminants contained DNA banding at $d = 1.720$ and 1.714, covering the range of intermediate-density DNA observed by Ledoux and Huart (1969). Next, barley seedlings known to be contaminated were used as experimental material in DNA feeding experiments, and this time, a prominent intermediate-density peak appeared (fig. 1.6).

Finally, Kleinhofs demonstrated that DNA from the intermediate peak hybridized specifically with DNA from the contaminating bacteria (named BC2Y) but not with that from the donor DNA (fig. 1.7). These results were later confirmed, extended and published in collaboration with researchers at the University of Washington in Seattle (Kleinhofs 1975; Kleinhofs et al. 1975).

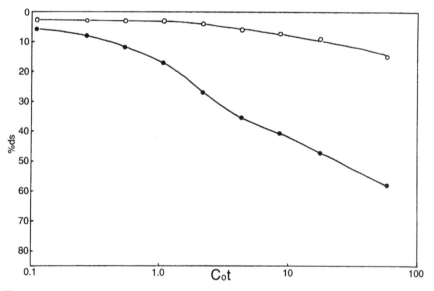

FIGURE 1.7.
Reassociation kinetics of "intermediate" DNA with contaminant and M. *lysodeikticus* donor DNAs. The intermediate-density peak from figure 1.6 was denatured and its reassociation kinetics studied in the presence of M. *lysodeikticus* DNA (*open circles*) or DNA isolated from the BC2Y bacterial contaminant (*closed circles*). It can be seen that M. *lysodeikticus* DNA does not accelerate the reassociation of intermediate-density DNA, whereas that from the bacterial contaminant does. (Adapted from Kleinhofs, A. 1975. DNA-hybridization studies of the fate of bacterial DNA in plants. In L. Ledoux, ed. *Genetic Manipulation with Plant Material*, pp. 461–477. New York: Plenum Press.)

More Visitors

In the meantime, Yasuo Hotta, who in 1970 had not detected any unusual DNA molecules in barley seeds incubated with heteropycnic DNA under physiological conditions, had secured funding for a prolonged stay in the Mol laboratory. This was late 1974, and he found the issue of foreign DNA integration and replication in barley completely moot, at least as far as intermediate-density DNA peaks were concerned. He thus decided to work with a simpler system: plant cells in tissue culture. This technique would provide much more homogeneous cell populations, and the problem of sterility could be addressed much more easily. Thus, *Arabidopsis* callus cultures (in which the cells grow in undifferentiated masses) were established and incubated with a variety of ^{3}H-labeled heteropycnic bacterial DNAs. No intermediate-density DNA peaks were ever detected with this system and ultrasonication just broadened the banding pattern without causing any heterogeneity in the distribution. Of course, not finding an intermediate peak did not necessarily mean that small but biologically significant portions of donor DNA could not have become integrated within the plant genome, and such low amounts of foreign DNA would have no effect on the buoyant density. Therefore, P_0t curves of labeled *Arabidopsis* DNA, using bacterial DNAs as drivers, were generated. No difference between control *Arabidopsis* DNA and that originating from bacterial DNA-treated cells was found. The sensitivity of these assays was such that a level as low as 3 percent donor DNA in *Arabidopsis* DNA should have been detected (Lurquin and Hotta 1975). This, in fact, still constituted a very significant quantity of potentially biologically active DNA, and if the donor DNA had consisted of a selectable cloned gene, it would have been possible to detect its expression. In those days, there were of course no cloned plant selectable markers and, as a matter of fact, nobody had any idea what these could possibly be.

News that the notion of easily detectable integration of foreign DNA in plants had been overturned had not yet spread very much in 1975. As a result, yet another visitor from the United States came to Mol that year to investigate. Clarence Kado was in the Department of Plant Pathology at the University of California, Davis, and had long had an interest in crown gall. He thought that if indeed it is *A. tumefaciens* DNA that causes crown gall transformation (this was not known at the time), then an *in vitro* DNA uptake system could be able to effect transformation, as claimed by Ledoux. Of course, this meant verifying whether or not the claims were correct. Not only was the uptake of labeled *A. tumefaciens* DNA in cultured tobacco cells studied, but also the possibility of this DNA's replication was also checked. The usual CsCl studies showed no intermediate-density DNA here either, and filter hybridization and thermal chromatography on hydroxylapatite after solution hybridization showed that there was no detectable level of foreign DNA in the incubated plant cells (Kado and Lurquin 1976). The results with tobacco cells were indistinguishable from those obtained with *Arabidopsis* cells.

By then, it seemed that the issue of the intermediate-density peak was no longer important in determining the fate of foreign DNA in plants, and a fresh look at the problem was necessary. A small step in that direction had been taken by Robb Schilperoort in Leiden, who was also interested in DNA uptake studies in relation to crown gall, which was his specialty. There, his student Frans Heyn had been working on the uptake of *A. tumefaciens* DNA by tobacco cells in suspension (Heyn and Schilperoort 1973). Basically, they had shown that DNA can bind to plant cells grown in liquid medium, and that a small proportion of the DNA remained associated with protoplasts from these cells after enzymatic digestion of the cell wall. This might imply that a small amount of donor DNA could cross the cell wall and bind to the cell membrane. They had seen no intermediate peaks in CsCl gradient either, but that was no longer so interesting. What was much more exciting was a question posed by Edward Cocking of Nottingham, United Kingdom, which figured in the commentary section at the end of the article (which was part of a symposium held in Versailles, France): "I would like to ask you why you have used cells and not isolated protoplasts in your studies on DNA uptake?" And indeed, why not use protoplasts? They could be prepared in large numbers and did not clump the way cells did; they were devoid of a rigid cell wall, a possible impediment to DNA uptake; they could be manipulated and plated; and if isolated from the right species, they would divide to form callus cells and could even be regenerated into plants. As it turns out, protoplasts would be instrumental in the development of techniques that finally resulted in successful direct gene transfer many years later (see chapter 4).

Meanwhile, the story of the intermediate-density peak in Ledoux's lab was nearing the end. The last act was played in 1975, when Mary-Dell Chilton of the University of Washington in Seattle, who had made a number of unsuccessful attempts to reproduce an intermediate-density DNA species in barley, requested a CsCl gradient fraction containing a putative hybrid between *M. lysodeikticus* and barley DNA actually generated by Ledoux himself in his own lab. Ledoux obliged by sending DNA from barley seeds that had been incubated with ^{3}H-DNA from *M. lysodeikticus* and then ^{14}C-thymidine. The purpose here was to study at once "integration" of the tritiated DNA and "replication" of the intermediate-density DNA claimed to be present in the sample. Supposedly, the seeds used in this experiment had been axenic. Unfortunately, when Mary-Dell Chilton rebanded this DNA in CsCl, she found that the ^{3}H and ^{14}C peaks did not coincide and were thus not present in the same molecular species. In fact, the [14]C "replicated" DNA banded at a density lighter than that of the ^{3}H-DNA, which banded near that of the donor. Reassociation kinetics showed that the ^{3}H-DNA was from *M. lysodeikticus*, whereas the DNA found in the ^{14}C-labeled peak showed homology with neither *M. lysodeikticus* DNA nor barley DNA. There is little doubt that the ^{14}C-labeled peak was again the result of microbial contamination that had gone undetected, and the ^{3}H-labeled peak was simply residual donor DNA. These data were never published, although they were disseminated by mail among a small group of scientists directly involved in "intermediate-density DNA" research.[1] This signaled the end of the intermediate DNA saga in barley.

Other Observations of Intermediate-Density DNA in Plants

Also in that year, the intermediate-density DNA bands seen in experiments with tomato [the material in which they were originally detected (Stroun et al. 1967)] were reinterpreted. Hanson and Chilton (1975) demonstrated the existence of a shoulder in the banding pattern of tomato DNA in CsCl gradients. This shoulder became more prominent when tomato shoots were incubated in the presence of unlabeled *A. tumefaciens* DNA and then labeled with ^{3}H-thymidine. This satellite DNA was purified by rebanding in CsCl and shown to have a density of 1.704—that is, intermediate between the densities of *A. tumefaciens* DNA ($d = 1.718$) and tomato DNA ($d = 1.692$). Further, after shearing by sonication or French pressure cell, this satellite was shown to separate into two bands, one of $d = 1.711$ and the other of $d = 1.699$! Even though the buoyant densities of these two peaks did not exactly match those of the donor and recipient DNAs, it is easy to see the parallel with experiments in which integration and replication of foreign DNA in tomato had been claimed. This intermediate-density satellite was analyzed by filter hybridization and reassociation kinetics and shown to contain no detectable homology with *A. tumefaciens* DNA, whereas roughly 50 percent homology would have been predicted by the integration/replication model of Ledoux. Further studies showed that this satellite is present in tomato DNA as a multicopy sequence, and based on its density the authors proposed that it might be mitochondrial DNA.

I also found a satellite peak, both in UV absorbance and radioactive labeling, in *Arabidopsis* DNA with $d = 1.706$ (Lurquin 1977), after rebanding twice the heavy side of a CsCl density gradient peak. These DNA preparations originated from callus cells cultivated under axenic conditions. The existence of this satellite had never been reported by Ledoux et al. (1971) in their controls, quite possibly because the specific radioactivity of their samples was low and so was the total input of DNA in their gradients. I made no attempts to study the behavior of this satellite after sonication and do not know whether it represents mitochondrial DNA or a nuclear satellite of DNA. Whatever its source, the existence of this molecular species raises nagging doubts about claims of integration and replication of foreign DNA in *Arabidopsis*, which were based in part on peak heterogeneity appearing after CsCl rebanding of heavy fractions.

Interesting but somewhat curious phenomena were reported by Vera Hemleben's group in Tübingen in 1975 and 1976 (Hemleben et al. 1975; Gradmann-Rebel 1976). The crucifer *Matthiola incana* was used by these investigators to study the uptake and fate of *homologous* DNA in young seedlings. To differentiate the endogenous DNA from the donor DNA, two types of experiments were done: In one series, the donor DNA was isolated from ^{3}H-adenine-labeled seedlings that had also been grown in the presence of 5-bromodeoxyuridine (BrdU). The incorporation of BrdU shifted the density of *Matthiola* DNA from 1.698 to 1.741. This DNA was then fed to normally grown *Matthiola* seedlings in the presence of unlabeled adenine to minimize incorporation of radiolabeled donor DNA breakdown products.

In the second series, *Matthiola* seedlings were grown in the absence of radioactive DNA precursors but in the presence of BrdU to render the endogenous DNA heavy. "Heavy" seedlings were then incubated with "light" radioactive donor DNA and unlabeled precursors, as described previously. After incubation, seedlings were treated with DNase to remove loosely bound donor DNA, and DNA was extracted and analyzed by CsCl gradient centrifugation. Results showed the presence of donor DNA in both cases (this was expected) and, more interestingly, the presence of a prominent, intermediate-density peak at $d = 1.717$ to 1.719 (fig. 1.8). Such an intermediate peak was not detected when the donor DNA had been depolymerized by sonication, or heat-denatured prior to incubation with seedlings. Further, when a chase period followed incubation with the donor DNA, exogenous DNA disappeared from the seedlings (again, this is expected as donor DNA is not anticipated to remain free and highly polymerized for very long), whereas the intermediate DNA fraction was stable.

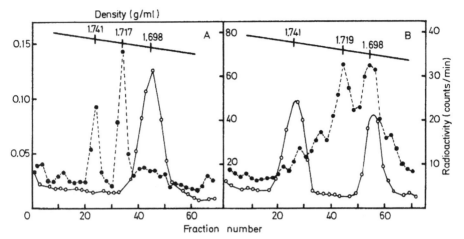

FIGURE 1.8.

"Intermediate-density" DNA in *Matthiola incana*. *Left: Matthiola* seedlings were incubated with radio- and density-labeled homologous DNA ($d = 1.741$), and their DNA was extracted and analyzed in a CsCl gradient. The intermediate-density DNA peak bands at $d = 1.717$. *Right:* Density-labeled *Matthiola* seedlings were incubated with "light," radiolabeled homologous DNA, and their DNA was extracted and analyzed in a CsCl gradient. The intermediate-density DNA bands at $d = 1.719$. *Open circles:* UV absorbance (light *Matthiola* DNA bands at $d = 1.698$, whereas heavy *Matthiola* DNA bands at $d = 1.741$). *Closed circles:* Radioactivity. (From Hemleben, V., N. Ermisch, D. Kimmich, et al. 1975. Studies on the fate of homologous DNA applied to seedlings of *Matthiola incana. Eur. J. Biochem.* 56:403–411, figure 2. Copyright Blackwell Science, Ltd. Reprinted by permission of the publisher.)

Finally, sonication of the intermediate-density peak generated two bands, one at the density of the donor DNA and one at the density of the intermediate. This was in fact a practically full reproduction of Ledoux's results, obtained here with homologous DNA. Autoradiographic studies also showed that at least a portion of the radioactivity resided in isolated nuclei. The authors interpreted their results as showing that donor DNA had been taken up and covalently integrated within the recipient genome. Here also, taken at face value, these results seemed very convincing, although most of the data were obtained with "light" seedlings incubated with "heavy" DNA. It is thus not known whether the "mirror image" effect holds true through the whole story. Was bacterial contamination involved? This is not known; the seeds were sterilized prior to treatment and the controls were clean. Was a satellite involved? We will see later that *Matthiola* seedlings, in other experiments, responded unexpectedly to heterologous DNA. Was a toxic effect (stress) of BrdU responsible for these observations?

Difficult to tell. A close study of the authors' data reveals that their "heavy" donor DNA had not homogeneously incorporated BrdU, as shown by a definite skewness and even prominent shoulders toward, unfortunately, the light side, in their donor DNA CsCl distributions. Could this uneven distribution of BrdU be responsible for some of the observed effects? Who knows? These results were never independently confirmed by others and thus remain intact as part of the "intermediate DNA" legacy.

Further experiments by the same group (Gradmann-Rebel et al. 1976) cast some doubt on the validity of their previous interpretation of the fate of exogenous DNA in *Matthiola*. This time, seedlings were incubated with radioactive T4 phage DNA. This DNA bands at $d = 1.694$ and is thus *lighter* than the DNA from the recipient ($d = 1.698$). Oddly, when *Matthiola* seedlings were incubated with T4 DNA, most of the radioactivity was found at the level of the host DNA, but a very significant portion was found at a density of 1.724, much *higher* than the densities of either the donor or the recipient. Controls without T4 DNA did not show an endogenous band at that density. This of course posed a major problem: How could DNA of low density become integrated within low-density host DNA and generate a high-density molecular species? The proposed answer was that phage T4 DNA had become specifically integrated within a high-density fraction of the *Matthiola* genome. To substantiate this point, the authors incubated seedlings with both radioactive T4 DNA and BrdU and found that both the bulk of *Matthiola* DNA and the new band (presumed to contain integrated T4 sequences) underwent the same density shift. This certainly suggested that the new band was DNA able to replicate. Upon sonication, the peak of density 1.724 yielded a narrow band at $d = 1.712$, which, according to the authors, corresponded to hybrid molecules still containing substantial amounts of T4 DNA integrated as short segments within the high-density component of *Matthiola* DNA. In a "classical" Ledoux type of intermediate DNA, one would have expected a peak at $d = 1.694$ (the density of the donor) and another one at the density of the "heavy" plant component. Therefore, sonication could not separate T4 sequences from the high-density component of the host DNA.

In addition, the nature of the $d = 1.724$ peak was studied by filter hybridization. Only those fractions banding at the level of the heavy peak showed a significant signal. However, the input radioactivity was so low that this signal was only about 17 counts per minute (cpm) above the blank and only about 10 cpm above some of the fractions where no hybridization was supposed to occur. It is very doubtful that these hybridization values are significant. Therefore, most of the evidence provided in this paper rests again on the CsCl density gradient centrifugation technique. Could it be that incubation of *Matthiola* seedlings with DNA resulted in the preferential synthesis of a natural satellite as it did in tomato (Hanson and Chilton 1975)? Was impure DNA at fault? These results cannot be interpreted easily, because no evidence for a high-density DNA component in the *Matthiola* genome was provided. We may never know the truth; there was no follow-up to these two studies.[2]

The odd observation described previously, that the amount of radioactivity found in intermediate DNA increased dramatically in barley and *Arabidopsis* seedlings, as well as in tomato shoots, after incubation with unlabeled bacterial DNA and then ^{3}H-thymidine, was not further explored by anyone. Since this increase is mirrored by a decrease of the endogenous DNA's specific radioactivity (at least in the case of barley), the following explanation can be offered: Assuming that the intermediate DNA is actually found inside contaminating bacteria, and given that bacterial cell division is much faster than plant cell division, it is reasonable to assume that ^{3}H-thymidine will be used much faster for bacterial DNA synthesis than for plant DNA synthesis. This would explain why more and more radioactivity was found over time in the intermediate DNA peak. But, then, why did the plant DNA specific radioactivity decrease so much in an actively growing organism such as germinating barley? It is quite possible that, as contaminating bacteria proliferate, they rapidly deplete the pool of available ^{3}H-thymidine and prevent its incorporation in the newly synthesized plant DNA.

Conclusions

Many studies published in the peer-reviewed scientific literature have been shown at a later time to be fundamentally erroneous. What can be said objectively about the results presented here? In other words, have the critics of the "intermediate-density DNA" actually proved their point? I would say they did so only in some cases, but then totally convincingly. Also, we now know that this DNA fraction can have multiple origins, not just bacterial contamination. Indeed, intermediate DNA peaks are observed under axenic conditions in the tomato example. There, it is shown that incubation with DNA, a possible cause of stress, led to preferential synthesis of an intermediate-density DNA that was a natural part of the tomato genome. Similarly, we saw earlier that physical and radiological stress will induce the synthesis of unusual DNA molecules in barley, although in the case of the experiments performed in Ledoux's laboratory, bacterial contamination was clearly responsible for the observed effects. However, foreign DNA integration and replication in *Arabidopsis* seedlings was never tested independently by other investigators under identical conditions, and in a sense this remains an open question. The same holds true for the *Matthiola* experiments. Also, published controls were always clean and, perhaps first and foremost, the density of the intermediate DNA was proportional to that of the donor DNA in several studies. This question was never addressed by others, although it provides great internal consistency to the integration/ replication model. Finally, the absence of intermediate-density DNA in *Chlamydomonas* could be attributed to some peculiarity of this organism, whereas it could be said that intermediate-density DNA was not seen in tobacco and *Arabidopsis* cells in tissue culture precisely because these were undifferentiated cells grown *in vitro*. Therefore, things are not that simple. But who would be willing to redo these old experiments with *Arabidopsis* and *Matthiola* seeds today, knowing that in the cases of tomato and barley, foreign DNA integration was not demonstrated and intermediate-density peaks were attributable to other phenomena?

Yet, we all know that these early reports of plant "transformation" are not sustainable, particularly in light of present-day results about how much foreign DNA actually gets integrated into plant cells and under what circumstances. For example, modern direct gene transfer technology invariably requires some kind of agent (e.g., electric discharge, accelerated metal particles) to force the donor DNA inside the recipient cells. Not a single technique in use today is based on the passive diffusion of donor DNA into intact plant cells. Further, plant cells exposed to foreign DNA have never been observed to integrate the equivalent of even a fraction of a bacterial genome in terms of mass. On the contrary, the copy number of integrated transgenes tends to be low, and very sensitive techniques are needed to detect the presence of these transgenes. This is in sharp contrast to the early results purporting to demonstrate foreign DNA integration and replication in plants. Therefore, taking all these results together, it is fair to conclude that intermediate-density peaks seen in barley, tomato, and *Arabidopsis* after incubation with heteropycnic bacterial DNA were either artifactual or, in some cases, ascribable to plant DNA with unexpected properties. In the end, the conclusion that bacterial and plant DNA could become joined *in vivo* relied on one single experimental observation: the splitting of the intermediate-density band under the action of ultrasound, which may very well have been caused by viscosity effects generated by highly polymerized density-marker DNAs present in the CsCl gradients. Poor quality of the plant

DNA samples may also have been a factor. The widely spread rumor that fraud was committed by one or several technicians of the Mol laboratory is difficult to sustain, as it has not been witnessed or reported by any of the numerous researchers, visiting and local, who spent considerable amounts of time in this laboratory. On the contrary, Ledoux's head technician, Raoul Huart, now deceased, freely opened his lab books for, and even collaborated with, those who eventually demonstrated that his research was flawed. If there was selection of the data to fit the model, the culprit(s) must be found elsewhere.

The controversy that revolved around the "intermediate DNA" and the refutation of its original meaning (integration of foreign DNA into the plant host genomes) raises deep questions regarding the nonexistence of seemingly new phenomena. In the scientific community, positive results, as long as they do not appear completely outlandish, will generally be accepted at face value. Failure by others to duplicate such results can easily be attributed to technical limitations, incompetence, or misunderstanding of experimental conditions. This argument was used repeatedly in the case of the "intermediate DNA" in plants. And indeed, it is immensely difficult to prove that something does not happen, let alone to have such "negative" results published. In the end, the only approach that worked in the main case described in this chapter was to spend time in the laboratory where the new claims were made (as A. Kleinhofs did) or to secure material from that laboratory and analyze it independently, as done by M.-D. Chilton. In both instances, the "intermediate DNA" proved to be artifactual.

Finally, it is ironic that horizontal (lateral) gene transfer, which is after all what was attempted by these early researchers, is now part of mainstream evolutionary thinking. Vilipended as the idea was in the late 1960s and early 1970s, it is now considered crucial, *mutatis mutandis*, in the understanding of the origin of the three domains of life, the bacteria, eukaryotes, and archaebacteria (Doolittle 1999). In that sense and in retrospect, it can be said that the idea of conducting DNA transfer experiments in these early days was scientifically visionary, albeit very premature.

Chapter 2

Genetic Experiments

ℰ

Cscl gradients and DNA hybridization experiments of the kind discussed in chapter 1 do not address whether integrated foreign DNA can be biologically expressed in the plant. If foreign genes were expressed, one would want to know whether they are inherited by the offspring of the transformed plants. For this, classical genetic experiments are needed, in which DNA encoding a trait easily scored is first fed to plants. If this foreign DNA is expressed in a detectable way in the new host, crosses can be made with this transgenic (transformed) plant to study transmission of the new trait and to test whether Mendel's principles are obeyed (or violated). This chapter summarizes the early claims of genetic transformation in plants and their subsequent refutation.

First Biological Effects Reported

Dieter Hess, now at Hohenheim University in Germany, seems to have been the first person to claim biological and genetic effects of exogenous DNA in plants. The fact that he published most of his results in the German language may have delayed their impact, but eventually his work came to be more widely recognized. Much credit for this gained notoriety should go to Bianchi and Walet-Foederer of the University of Amsterdam, who published in English an extensive summary of Hess's work and critiqued it (Bianchi and Walet-Foederer 1974). As their own article was published in a cryptic Dutch journal, it is mostly through the circulation of reprints that scientists learned about this new controversy.

Hess's early transformation work dealt with the effects of exogenous DNA on flower color in *Petunia hybrida* and was published between 1969 and 1972. Only his last paper on the subject will be cited here (Hess 1972). Petunias had long been an object of interest among geneticists, thanks to one major and easily visible characteristic, flower color, and to the existence of mutations affecting this color. This boon is also a bane; flower color, even in albino mutants, is quite sensitive to environmental conditions, and homozygous recessive mutants impaired in synthesis of anthocyanin (a class of pigments found in flowers) can sometimes show light coloration. Somatic mutations (that is, mutations not in the germ line and hence not heritable) also exist and can be responsible for sectorial flower pigmentation in the midst of a colorless background. In addition, anthocyanin synthesis depends on a complex pathway under the control of many genes. Thus, in Hess's experiments, eight-day-old seedlings of a white-flowering petunia mutant were treated with DNA extracted from a red-flowering strain, allowed to grow, and then examined for flower color. Any individuals having picked up the donor DNA would be expected to carry red flowers, an easily distinguishable trait. This experimental protocol was thus simple and straightforward, but the results were not. In a first series of experiments, Hess reported that up to 15 percent of the DNA-treated seedlings gave flowers exhibiting reddish to red (four individuals in the latter category) color. Why was there a variation in expressivity? No explanation was offered. Much more troubling was the fact that up to 9 percent of the seedlings treated with DNA from the white flowering mutant were seen to exhibit faint red color. Thus, instead of an all or nothing type of effect, there was a gradation of colors and the potential for subjective interpretation. Further, when these experiments were repeated in a totally different location in Germany, the "transformation" frequency dropped to 0.06 percent! Although such enormous variation between different experiments could easily be attributed to environmental factors,[1] this meant that it would be extremely difficult to verify and duplicate these results independently.

Hess also observed that the four dark-red-flowering "transformants" were homozygous for the trait, in that no white segregants appeared in the F2 generation after selfing (see the glossary for a definition of selfing and other terms). This also implies that the added pigment genes (the transgenes) must have been tightly linked to the original mutant alleles, perhaps even replacing them. Bianchi and Walet-Foederer (1974) pointed out that, because in *Petunia* there are in the apex at least three cells that give rise to the epidermis, the incorporation of the two red alleles must have occurred at least three times to account for the existence of uniformly red flowers. This, they said, would be an extremely unlikely event.

Now, what about the plants bearing flowers with faint coloration? These came in two types: flowers showing faint pigmentation throughout and flowers showing mostly faint but sometimes strong pigmentation along the veins of the corolla. Hess showed that they were heterozygous for the "added" trait by selfing and backcrossing to the original white-flowered mutant. This then suggests that, in this interpretation, the red allele is semi-dominant to white, a piece of information that was not given by the author. Or, as Bianchi and Walet-Foederer (1974) pointed out again, in their experience, environmental factors could be responsible for subtle phenotypic variations such as those dealt with by Hess. Further, they demonstrated that pigmentation of corolla veins can be the product of somatic mutations and that, as expected, selfing of plants carrying such mutations produced no corolla vein pigmentation in the offspring, as also observed by Hess. Finally, they showed that automatically assuming that an effect of DNA on flower color should necessarily be inheritable was a fallacy; they demonstrated that the eight-day-old petunia seedlings used by Hess already had a well-differentiated shoot apex, meaning that an effect of DNA (either as transformation, mutagenic effect, or other) could possibly have been limited to epidermal cells without altering future gametes, whose primordia are distinct from epidermis primordia. In summary, these two authors simply did not think that Hess's conclusions were supported by his data.

What lessons can we draw from this controversy? The one clear conclusion is that flower color in *Petunia* is far from being an ideal system for this type of experiment for at least three reasons. First, the phenotype is much too sensitive to the influence of the environment; for example, changes in lighting and pH are known to alter flower color in this organism. Second, it is very difficult to categorize individuals that show a variety of shades of red, as these judgments are necessarily subjective. Third, the "background" of somatic mutations and the appearance of faintly colored flowers in the controls treated with mutant DNA make the whole situation quasi-impossible to interpret. How could changes brought about by DNA be distinguished from normal variation in flower color?

In genetic jargon, mutations that do not result in clear-cut phenotypes (as in the preceding example, where albino mutants can sometimes produce pigments) are often called leaky. These mutations are not necessarily intractable, but they require sustained genetic segregation analyses to show that they conform to Mendel's laws. Further, the elucidation of the interactions between the environment and gene expression in complex organisms is still in its infancy. In brief, the material chosen by Hess was much too complicated to yield simple answers in spite of the apparent simplicity of the question posed. I will show later how other leaky mutants created havoc in the interpretation of transformation experiments.

Finally, because plants are multicellular organisms and only a tiny minority of these cells will eventually form the germline, it is unlikely that transformation events would target these cells preferentially in a growing plant, or even a seed, and lead to transformed progeny. This argument applies equally to the example that will be developed next.

Work with *Arabidopsis* Thiaminless Mutants

After this little foray into Germany, we now return to a familiar place in Belgium: Mol. The year is 1971, and Ledoux and collaborators have just published their extensive work on the integration and replication of bacterial DNA in *Arabidopsis* (Ledoux et al. 1971), in which they announce that genetic experiments with this organism are under way. These were of course an absolute necessity: Without a biological effect, plants containing large amounts of bacterial DNA would be little more than laboratory curiosities. It turns out that selecting *Arabidopsis* as experimental material was the only possible choice, because the only known and more or less well characterized biochemical plant mutants belonged to this species. I have already explained that *Arabidopsis* has a short life cycle, can be grown under well-controlled conditions, and, thanks to its very small size, can be propagated *en masse*. All these features make it an ideal organism for genetic experiments where large numbers of progeny must be examined.

The purpose of these genetic experiments was to determine whether bacterial DNA could complement (correct) mutations in the thiamin (vitamin B1) pathway in *Arabidopsis*.[2] Four different types of mutations in this pathway were known at the time; they are called *tz*, *py*, *th-1*, and *th-2*. Of the four, only the first three are lethal; the *th-2* mutant is viable. Thiamin is composed of two chemical moieties, a pyrimidine ring and a sulfur-containing thiazole ring, linked by a methylene bridge. The *py* mutants are unable to synthesize the pyrimidine but will survive if this compound is fed to them, either by spraying the seedlings or by including it in a growth medium solidified with agar. Similarly, the *tz* mutants are incapable of synthesizing thiazole but can be rescued by application of this compound. Finally, *th-1* mutants are unable to link the pyrimidine and thiazole rings together but will survive in the presence of added thiamin. Of course, both *py* and *tz* mutants can be rescued by thiamin, too. Figure 2.1 illustrates these three enzymatic blocks.

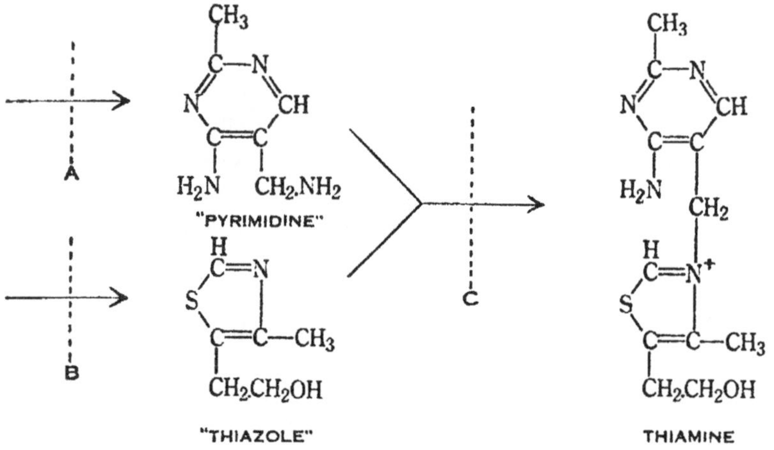

FIGURE 2.1.
Synthetic pathway of thiamin in plants. Mutant blocks are indicated by A, B, and C. (Adapted from Langridge, J., and R. D. Brock. 1961. A thiamine-requiring mutant of the tomato. *Australian J. Biol. Sci.* 14:66–69.)

The *th-2* mutants are more mysterious: They are viable in the absence of added thiamin, pyrimidine, or thiazole; they are often variegated (leaves showing patches of cells containing less chlorophyll); and variegation is increased while growth is depressed in the presence of glucose at low temperature. The other mutants and the wild type do not show these symptoms. The *th-2* mutants are thought to be mutated in a function that regulates the activity of the other genes involved in thiamin synthesis. These four types of mutations are nonallelic—that is, these genes are independent loci.

Ledoux and coworkers set out to determine whether DNAs of bacterial origin could complement (correct) thiamineless mutations in *Arabidopsis*. Given the seemingly successful results obtained with foreign DNA uptake/integration and replication experiments, this was a reasonable thing to do. Wisely, they did not use *th-2* mutants, because their viability would complicate data interpretation. The logic behind these experiments was that bacteria do use thiamin for growth and, like plants, are able to synthesize it, presumably via a similar pathway. In fact, this was shown to be the case in *E. coli*. Thus, if bacterial DNA molecules containing functional genes for thiamin synthesis are taken up, integrated and replicated in the *Arabidopsis* genome, and expressed, they might be able to correct thiamin deficiencies present in the mutants and render them able to survive and grow, independent of an outside supply of pyrimidine, thiazole, or thiamin itself.

The Ledoux team was apparently successful in correcting a variety of thiaminless mutants, and the work was published in *Nature* in 1974 (Ledoux et al. 1974). This is what they did and found. They incubated thiaminless mutant seeds with DNA extracted from a variety of bacterial sources (*A. tumefaciens, E. coli, B. subtilis, M. lysodeikticus,* and *Streptomyces coelicolor*), all able to synthesize thiamin normally or unable to do so by mutation (such as *E. coli* P678 *thiA,* blocked in the synthesis of thiazole). Further, other seeds were incubated in the presence of phage T7 DNA or phage 2C DNA, both known to not contain any thiamin genes. Control seeds were incubated in the presence of 0.01 mol/L NaCl without DNA at all. Treated seeds, which by the end of the incubation period had germinated, were transplanted into growth medium without any thiamin or its precursors, to select for transformants.[3] About 15,000 seeds underwent a variety of treatments, providing the following results: All controls were negative—that is, mutant seeds treated with saline, phage T7, and phage 2C DNA never gave rise to adult plants, whereas DNA preparations from all bacterial sources were able to correct deficiencies in all the mutants used, albeit with varying degrees of success depending on the source of the DNA and the type of mutant considered. For example, *E. coli* DNA was the most effective and *S. coelicolor* DNA was the least. DNA from the thiazole-minus *E. coli* mutant (P678) was unable to correct the *Arabidopsis tz* mutant but did correct a *py* mutant, as one would expect. Overall transformation (correction) frequencies were 0.7 percent for the *th-1* mutant, 0.85 percent for *tz,* and 0.49 percent for *py.* These frequencies are quite high, meaning that others should have been able to reproduce these results without having to deal with inordinate numbers of seeds.

Unfortunately, this was not the case. In a short report, Feenstra's group in The Netherlands (the provider of one of the *py* and one of the *th-1* mutants used by Ledoux) was unable to isolate a single corrected mutant from 2,000 treated *py* mutants (Feenstra et al. 1973). To my knowledge, this was the only published serious attempt to duplicate Ledoux's findings.

The next step was to characterize these corrected mutants, and this was done by selfing and outcrossing (see appendix 3). The purpose was to determine whether the corrected plants were hemizygous (having only a single copy of the correcting gene) or homozygous for the correction. We know today that transforming DNA integrates randomly within recipient genomes. Thus, in retrospect, one would have expected the same in Ledoux's *Arabidopsis* correction experiments. This was not at all the case. In fact, selfing experiments and testcrosses ruled out this possibility, as no segregation of the mutant locus was observed at all. At the very least, one had to think that the correcting DNA was integrated very close to these loci, and this at least twice, because the corrected plants behaved like true-breeding homozygous dominant individuals. The analogy with Hess's previous results is obvious and was taken at the time as strongly supporting evidence. This situation, absence of segregation by selfing and testcrossing, was explained away by a putative meiotic drive effect in which only corrected gametes survived and/or were fertile.

However, things got more complicated. Segregation of the mutant allele *did* occur in the F2 and F3 of testcrosses (and crosses with the wild-type) although no segregation occurred in the F1. Additionally, these advanced generations showed the appearance of new individuals, called leaky mutants because they showed patches of white discoloration in leaves and stems and responded to thiamin or its precursors for growth. This was taken as evidence that they were heterozygous for the added bacterial allele. However, the numbers observed in the F2 and F3 generations of selfed leaky mutants did not correspond to Mendelian ratios; there should have been 25 percent lethals (homozygous recessives) when only 8 percent to 10 percent were recovered, and 50 percent of the leaky mutants (if indeed they were heterozygotes) should have been observed when in fact 65 percent to 70 percent were seen. Nevertheless, the conclusion was that the correcting bacterial genes had not replaced the original mutant loci since the latter reappeared in F2 and F3. As seen in appendix 3, assuming as claimed by Ledoux et al. (1974) that the correcting allele is distinct from the original mutant locus, the two must still be very tightly linked, a really unlikely occurrence, there being no significant homology between donor DNA and *Arabidopsis* DNA.

How were the disappearance of the mutant allele in F1 and its reappearance in F2 and F3, as well as the skewed phenotypic ratios, explained? It was hypothesized that the added trait was somewhat unstable and could be lost upon chromosome pairing during meiosis. Why this did not happen in the F1 of testcrosses or upon selfing remained a mystery. At this point, it is very difficult to draw any solid conclusions regarding the validity of these interpretations; given the unorthodox behavior of the alleles in crosses, explained by unsubstantiated meiotic drive and gene instability, caution is necessary. Such concerns were expressed in the "News and Views" section of the issue of *Nature* that published the Ledoux et al. (1974) article (Anonymous 1974).

At that time, Monsanto Company had started showing interest in the prospect of genetically engineering plants. This start was very modest but significant enough to send one of their scientists to Mol to test the waters.

Thus, Harold Weingarten spent a few days in the lab and requested some of the thiaminless mutant-corrected seeds. Why all this? Very simply, Monsanto had tried to reproduce the *Arabidopsis* correction results and had failed. Weingarten had thus come to Mol to understand why, but he did not stay long enough to do experiments. Eventually, several corrected seed stocks from the Ledoux lab reached him, and, in collaboration with the Rédei group at the University of Missouri, he repeated and extended the genetic experiments of Ledoux et al. (1974), without however trying again to produce new corrected individuals.

A year or so after all these corrected seed stocks had reached the United States, in 1976, the Biological Research Center of the Hungarian Academy of Sciences in Szeged had organized a UNESCO-sponsored summer course on plant cell genetics. Rédei, one the speakers, surprised everybody by announcing in his talk that Ledoux's corrected thiaminless mutants had nothing to do with DNA-mediated transformation: They were likely to be the result of a mechanical contamination of seed stocks! This is what they found (Rédei et al. 1976). The chromosomal locations of the *th-1*, *tz*, *py*, and *th-2* loci were well known and, since the results of Ledoux et al. implied tight linkage between the original mutation and the correcting gene, it was possible to map the correction with great accuracy, something Ledoux had not done.

Three corrected seed stocks, two from *py* mutants and one from a *tz* mutant, were genetically analyzed with the following results. First, all three groups were shown to be allelic, meaning that all three corrected loci were identical. This was very surprising because *py* and *tz* are not even on the same chromosome. Further, they could not detect the presence of the original mutations, directly in contradiction with the results published by Ledoux et al. (1974), and the correction in the *py* mutants was not flanked by the *er* and *as* markers (used here in three-point-crosses), as should have been the case. The last blow was delivered when trisomic analysis revealed that all the corrections were located on chromosome 5, near the centromere, where the *th-2* locus is situated. Finally, allelism tests with a genuine *th-2* mutant showed that indeed all the corrections were allelic to *th-2*! Physiological tests (nutritional requirements, temperature sensitivity, and dormancy) corroborated the interpretation that all the corrected types were indistinguishable from a *th-2* mutation! It will be recalled that the *th-2* mutant is indeed leaky (not lethal) and variegated, and thus it resembles very much the supposedly heterozygous plants found by Ledoux et al. in their F2 and F3 generations. Rédei et al. (1976) thus concluded that all the corrected types sent them by Ledoux were *th-2* mechanical contaminants in his correction experiments. Now, Rédei was at the time *the* expert on *Arabidopsis* genetics, and his concerns had to be addressed. Thus, following Redei's critique of his genetic data, Ledoux, with help from others, provided a more detailed genetic analysis of his own material. This work took approximately two years. What follows is a summary of the Ledoux group findings after the Szeged announcement. The results of physiological experiments published by Rédei et al. (1976) were partially reproduced, but not quite exactly; the effect of low temperature and glucose was reproduced with three of the four corrected lines provided earlier to Rédei but not with the fourth one. However, response to thiamin precursors was different in the corrected lines from that of two *th-2* mutants. Rédei's results were confirmed in crosses between a *th-2* mutant and a single corrected strain. Control crosses between *th-2* mutants and a *tz* mutant showed that complementation between these two loci was not complete, an unexpected result, which, according to Ledoux, made the interpretation of any crosses involving *th-2* difficult. Also, three-point crosses were made between the triple mutant *as py er* and two corrected lines, one of which had been previously mapped by Rédei et al. (1976) and diagnosed as being a *th-2* contaminant. In both cases, the correction was found to be tightly linked to the original *py* mutation, thus contradicting Rédei in one case. Strangely, however, the corrected line not analyzed by Rédei yielded no variegated plants in F2, whereas the previously studied one did. One is thus led to conclude that sometimes the correction (if it was a correction) was unstable and sometimes it was not. In short, there was some agreement with Rédei's findings, but far from enough to retract earlier conclusions. These results were published in 1979 (Ledoux et al. 1979) and never elicited any response. In the meantime, two extensive reviews had been published (Kleinhofs and Behki 1977; Lurquin 1977), concluding that there existed no sustainable evidence supporting the claims of integration, replication, and genetic activity of foreign DNA in plants. These conclusions were also Edward Cocking's, who wrote a courteous but very critical editorial in *Nature* (Cocking 1977) on Ledoux's collective work, while recognizing his importance as a pioneer.

Seed stock contamination seems indeed to have been a problem in the Mol laboratory, as demonstrated in figure 2.2. In this experiment, *th-1* and *tz* mutant seeds obtained from Ledoux's collection were allowed to germinate on B5 medium either without added thiamin or supplemented with thiamin, the pyrimidine compound or the thiazole compound, also obtained from that lab. B5 medium contains phytohormones that prevent the formation of green plants and force the cells to grow as undifferentiated masses, or calli. It can be seen that the *th-1* seeds (THIA⁻) grew only on medium

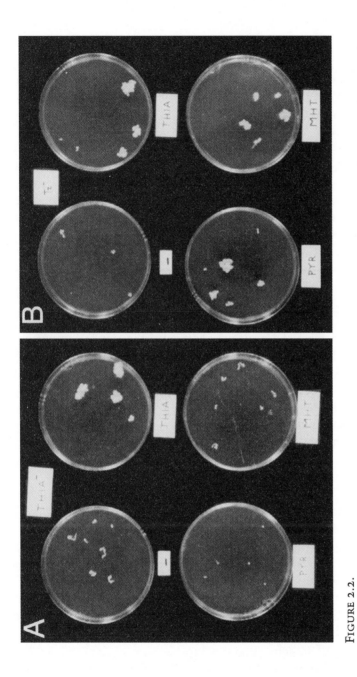

FIGURE 2.2.

A: Growth of *Arabidopsis th-1* (THIA⁻) mutant seeds on unsupplemented B5 medium (–), or on medium supplemented with thiamin (THIA), pyrimidine (PYR), or 5-β-hydroxymethyl-4-methyl thiazole (MHT, or thiazole). B: The same experiment conducted with *Arabidopsis tz* (TZ⁻) mutants.

supplemented with thiamin, indicating the absence of contaminants in that sample. However, the *tz* mutant seeds (TZ⁻) responded unexpectedly: Whereas most of these seeds developed in the presence of added thiamin and thiazole (MHT), as expected, at least one out of six seeds also developed vigorously on the pyrimidine compound. This can be explained only by the presence of *py* mutant or wild-type seeds in the *tz* stock. Similar results were obtained with the *py* mutant.

Here again, data selection has been repeatedly invoked by the scientific community to explain the published results of Ledoux's group. I suppose this was a natural reaction, given the controversial nature of the claims. Yet, without access to the original notes from this laboratory and in the absence of a whistleblower, it is of course impossible to determine what really happened. The heavy cloak of doubt remains.

E. coli β-Galactosidase Expression in Plant Cells

As we have seen, most of the positive claims about foreign DNA uptake and expression in plants had come from only two labs, Ledoux's and Hess's, and they, in some ways, had published mutually supportive results. However, other groups have made similar claims with other plant systems.

In 1971, Carl Merril and associates, then at the National Institutes of Health, Bethesda, had published in *Nature* an exciting article in which the claim was made that human galactosemic cells (galactosemia is a serious genetic disease in humans) deficient in galactose-1-P-uridyltransferase became able to survive in the presence of galactose (which normally is toxic to them) after incubation in the presence of high titers of λpgal or its DNA (Merril et al. 1971). Lambda pgal is a transducing phage (for explanations, see later) carrying the *E. coli* galactose operon, including the transferase gene (T), which performs a function analogous to that of the human gene that is deficient in galactosemic patients. Bacteriophage-specific mRNA was detected in the λpgal-treated human cells by DNA-RNA filter hybridization, and extracts from these cells converted uridine diphosphate–galactose into galactose-1-P. Control experiments with λpgal carrying a mutation in the transferase gene were negative. Even though no evidence was given that the transferase was of *E. coli* origin, circumstantial evidence was good and strong. The problem was that these results turned out to be irreproducible a few years later.[4]

What does this have to do with plants? Well, others followed in the footsteps of Merril and collaborators and developed similar systems, based on phages λ and φ80 (another lambdoid phage) in plants. If transducing bacteriophages worked in human cells, perhaps they would too in plant cells. Also, plant cells release DNase, which of course will destroy exogenous DNA. DNA present inside phage heads is protected from the action of DNase because it is wrapped in a protein coat. Thus, if phage particles can be taken up by plant cells, their DNA will be protected from DNase injury. The flip side of this is, of course, that the DNA must become unwrapped to be transcribed. This occurs readily in infected *E. coli*, but there was no evidence that plant cells (or human cells for that matter) could do this. This hypothesis was tested by Doy and his associates in Australia, who investigated the ability of λ and φ80 to transduce *E. coli* genes into cultured plant cells. Their efforts were successful until strongly questioned about a year later. They even coined a new term, transgenosis, to name their freshly discovered phenomenon. The word did not stick, but it almost did (often, the genetic transformation of higher organisms is called transgenesis).

Let us now study the experimental protocols and results of Doy et al. (1973a). They started from the premise that galactose and lactose cannot be used as carbon sources by plant cells in culture, unlike, of course, glucose. Now, it is well known that *E. coli* is perfectly able to grow on these sugars thanks to its *gal* and *lac* operons. The structural genes of the *gal* operon code for functions that convert galactose into glucose derivatives, and the *lac* operon–encoded enzymes cleave lactose into glucose and galactose. The key enzyme in the latter case is β-galactosidase, coded for by the *lac z* gene. Thus, if such operons (or their relevant structural genes) could be expressed in plant cells, these might conceivably become able to grow on lactose or galactose.

Lambdoid phages such as λ and φ80 are able, after infection, to integrate their DNA into the genome of their host, *E. coli*. This phenomenon is called lysogeny, and it is reversible in nature. That is, the phage genome can excise itself out of the host's DNA and become free again. This mechanism is normally extremely precise, and the phage DNA is cleanly excised. Sometimes, however, the excision mechanism makes an error and cuts DNA outside of the phage DNA borders. When that occurs, neighboring *E. coli* DNA will remain linked to the phage DNA, and, when the latter is encapsulated by phage proteins, this *E. coli* DNA is physically packaged within the phage head, along with the phage DNA. This phage now becomes able to *transduce E. coli* genes to other *E. coli* cells after infecting them. Because phages λ and φ80 have a propensity to integrate near the *lac* and *gal* operons in the *E. coli* genome, these are the operons that will be found in transducing strains of these two phages. This process is illustrated in figure 2.3. Such

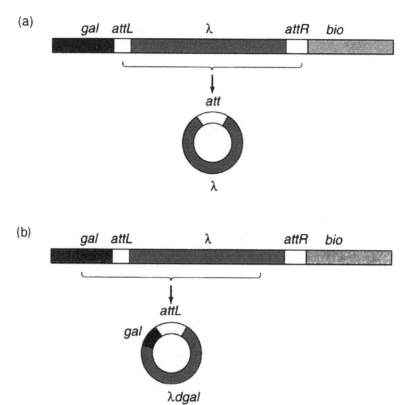

FIGURE 2.3.

Creation of a *gal*-transducing bacteriophage λ. (*a*) Normal excision of the integrated λ genome via clipping in the attachment (*att*) sites; (*b*) Excision occurring too far to the left, resulting in the inclusion of the *E. coli gal* operon and formation of λ*dgal*. This phage is defective in replication (hence *d*) because it lacks a portion of the λ right arm, where replication functions are located. (Adapted from Weaver R. F., and P. W. Hedrick. 1991. *Basic Genetics*. Oxford: Wm. C. Brown.)

transducing phage strains are readily available from many labs in the world. They are in effect cloning the flanking DNA, making possible the production of a high concentration of DNA that is a single gene (or region) from *E. coli*.

Thus, the logic of the experiment was to administer these transducing phages by simply adding aqueous suspensions of phage particles to plant cells in culture, to see whether these cells would become able to survive and grow in the presence of lactose or galactose. The result of these experiments was a resounding yes: Tomato and *Arabidopsis* cells in tissue culture were able to grow, although very slowly, on lactose or galactose if first exposed to high titers of transducing phages containing a functional galactose operon or lactose operon, including *lac z* (which, you recall, codes for β-galactosidase, the enzyme that cleaves lactose into glucose and galactose). The evidence provided included weighing the plant cells at various times after transfer onto the unusable sugars. As a control, phages with a mutation in the T (transferase) gene or phages λ and φ80 without the *gal* or *lac* operon were administered to the plant cells. All the untreated control plant cells, or those treated with phage that lacked the *lac* operon or *gal* operon, died within three weeks. In contrast, the ones treated with the *gal* or *lac* operon lived and proliferated, albeit slowly, for up to 18 weeks on galactose or lactose as sole carbon sources.

What's more, the authors demonstrated that the galactosidase activity present in extracts of cells proliferating on lactose could be specifically protected against heat denaturation by prior incubation with an antibody made against purified *E. coli* β-galactosidase. Some endogenous galactosidase activity was found in control cells, but levels were only about 1/25th of those found in treated calli; furthermore, this endogenous activity was not protected against heat by the specific antibody. This indicated that the endogenous activity was not structurally (and hence genetically) related to *E. coli* β-galactosidase. These observations raised a number of interesting questions: (1) If phage particles are taken up by plant cells, what is the mechanism in these cells that allows uncoating of the phage DNA to allow its transcription by plant RNA polymerase(s)? (2) How do plant RNA polymerases recognize and "read" the promoters of the *lac* and *gal* operons, which are of course "typical" *E. coli* promoters? (3) All three genes of the *gal* operon, the kinase, the transferase, and the epimerase (K, T, and E), must be expressed to convert galactose into glucose-1-phosphate, which is the sugar usable by plant cells. This then means that the whole *gal* operon must be transcribed into a polycistronic mRNA in plant cells and subsequently translated, unless plant cells provide some of these functions. Is this happening? (4) As the plant cells proliferate, the number of phage particles per plant cell must decrease, unless the phage (or its DNA) can replicate in this totally foreign environment. Does a sustained biological effect then mean that bacteriophage can replicate in plant cells?

These questions have not been answered in a positive manner to date, except for number 4. Indeed, two curious reports by Sander (1964, 1967) claimed to demonstrate replication and encapsulation of *E. coli* phage fd DNA (which has a single-stranded DNA genome) in tobacco leaves after inoculation with this phage's DNA. Intact phage fd, on the other hand, when inoculated into tobacco, did not multiply. Moreover, replication of fd DNA in tobacco resulted in host range reduction similar to that encountered when the phage is propagated in *E. coli* K12 versus *E. coli* B! This then suggested that tobacco and *E. coli* have comparable DNA restriction-modification systems! Finally, all these phenomena were observable only in winter; no other season would do. Such results have never been duplicated. We will see later that there is no evidence for phage multiplication in plant cells; on the contrary, there is evidence against it.

Doy et al. (1973a) made one more claim. Nonsense mutations, which turn a sense codon into a translational stop codon, usually prevent gene expression through premature termination of mRNA translation. Some tRNA anticodons, through mutation also, can suppress (correct) a nonsense mutation in a gene by recognizing the nonsense (stop) codon as a sense one. For example, the creation of a UAG codon through mutation in a gene is a nonsense mutation. It turns out that one of the tyrosine tRNAs has an AUG anticodon, which, through mutation, can be turned into an AUC. This AUC can base-pair with the UAG stop codon, with the result that a tyrosine will be incorporated at the level of the stop. The nonsense mutation is thus suppressed. This system was unraveled in *E. coli*, and often such suppressed mutants do not grow as well as nonmutated strains. This is in part because the mutated tRNA^{tyr} will recognize natural UAG stop codons at the end of a coding sequence, and normal translation termination will not occur. On the basis of this, Doy's team inoculated tomato and *Arabidopsis* cells with transducing $\phi80supF^+$, which carries a mutated tRNA^{tyr} able to recognize UAG stop codons. The cells stopped multiplying (fig. 2.4). This indicated that reading UAG as tyrosine rather than stop interfered with protein synthesis so badly that inoculated cells could not survive. Evidently, this also meant that plant RNA polymerase(s) were able to correctly transcribe the *E. coli* tRNA^{tyr} gene and, that the plant cells' aminoacyl-tRNA synthetases were able to correctly load tyrosine onto this heterologous tRNA. There is no evidence for any of this. The nagging thing in Doy's work, however, was that sometimes control phages caused cell death, which they were not supposed to do, and sometimes cells did not respond to the *supF*$^+$ gene.

In a follow-up study, Doy et al. (1973b) showed that mixed infections of plant cells with λp*gal* and $\phi80$p*lac* led to significantly higher levels of β-galactosidase in cells grown on lactose. One wonders why; no cooperative effect is expected here. Also, the time course of the appearance of galactosidase was given; it took about 65 to 75 days for the activity to peak (depending on the size of the phage inoculum) and then decrease significantly. A more interesting experiment would have been to check for syner-

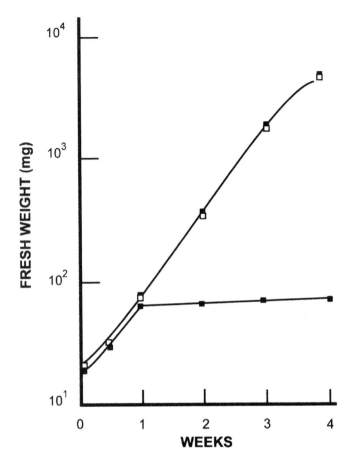

FIGURE 2.4.
Inhibition of growth effect of phage φ80 *supF*⁺ on tomato callus cells.
Closed inverted triangles: Control without phage. *Open squares*: Control with φ80. *Closed squares*: Incubated with φ80 *supF*⁺. (Adapted from Doy, C. H., P. M. Gresshoff, and B. G. Rolfe. 1973a. Biological and molecular evidence for the transgenosis of genes from bacteria to plant cells. *Proc. Nat. Acad. Sci. USA* 70:723–726.)

gistic effects of the two operons on growth on lactose medium; indeed, in the presence of lactose, the galactosidase should produce glucose plus galactose, the latter then being converted by the products of the *gal* operon into glucose. There would then be more glucose available, and growth would be promoted accordingly. This should not happen on galactose-containing medium. Such experiments were not reported.

Clearly, the results of Doy et al. (1973a, 1973b) had dramatic implications for the genetic modifications of plants, even though they raised more questions than they answered and they provided incomplete experimental evidence. For example, heat protection of galactosidase activity by an antibody is still indirect proof; it would have been better to be able to see an actual immunochemical reaction of plant extracts with purified antibody raised against E. coli β-galactosidase. This was not done, or at least it was not shown.

Exciting and unusual claims such as these are sure to attract attention, critical or otherwise, from others working in this field. This is exactly what happened, within a matter of months. This time, however, instead of a summary repudiation of the results of Doy et al., confirmation, with a twist, was published by Johnson et al. (1975) from the United Kingdom. Sycamore cells were shown to be unable to grow on lactose-containing medium, but when they were treated with transducing λplac5 (a transducing phage containing a functional lac z gene like that used by Doy et al. 1973a), they did grow, and all controls were negative. Was this finally the long-awaited independent verification that had been so cruelly lacking in this whole field? Not quite, because Johnson et al. were unable to detect the E. coli enzyme immunologically and, moreover, they felt uncomfortable with the levels of endogenous glycosidases present in control cultures. Nevertheless, a gross phenotypic effect had finally been reproduced. However, was E. coli β-galactosidase really responsible for this new phenotype, or did the latter have nothing to do with transducing phages whatsoever? The answers were provided by this same group, and they were no and yes, respectively (Smith et al. 1975).

Basically, the British team showed that cells grown in the presence of lactose did display a transient burst of galactosidase activity (as reported by Doy et al. 1973b) and that peak levels were approximately twice as high in cells treated with λplac5. However, they also showed that transfer of cells slowly growing in lactose medium, to fresh lactose medium, by subculturing over a period of several weeks, actually resulted in a decrease in cell numbers, not an increase. At best then, the transducing phage would have been transiently active, not stably incorporated and expressed. Also, phage titers decreased in the presence of plant cells, ruling out phage multiplication. Finally, a specific inhibitor of the E. coli enzyme failed to inhibit galactosidase activity in λplac5-treated cells. Further, they could not detect E. coli β-galactosidase immunologically, nor were they able to demonstrate protection against heat by a specific antibody. Their conclusion was that there was no evidence for the expression of E. coli genes in plant cells treated with transducing bacteriophages, in disagreement with Doy et al. (1973a).

It may never be known why plant galactosidases are stimulated by incubation with transducing phages; work with these phages stopped in these two labs but continued, this time using green plants, for another decade in Dieter Hess's. Further, the effect of a suppressor tRNA gene on cell growth was never confirmed or refuted. The mystery remains.

Hess in the meantime had turned his attention to pollen grains as carriers for exogenous genetic information. This was a good idea, because of course pollen grains are very much part of the genetic process and might very well bind DNA, phage particles, and so forth (plant cells can bind nonspecifically all sorts of macromolecules). Thus, uptake would perhaps not even be needed here; pollen grains could act as passive carriers of externally bound DNA or phages, which could possibly remain stuck to the elongating pollen tube during fertilization. This idea was first tested by treating petunia pollen grains with *lac* transducing phages and using them to fertilize flowers (Hess 1978). Progeny seeds were then sown on medium containing lactose and observed for growth. This was actually a strange approach; doing these experiments with tissue culture cells made sense, since these cells need an external carbon source for growth in the absence of photosynthesis, but green plants as used by Hess do photosynthesize and do produce the sugars they need. What then was he measuring? Lactose toxicity? Something else? Whatever the case may be, plantlets that had one parent that originated from phage-treated pollen seemed to tolerate, in terms of growth, lactose better than the controls. There were wide variations, however, and in two sets of experiments, the controls actually grew better than the λp*lac*-treated plantlets. This ability to grow on lactose was observed over five generations obtained by selfing of the plants that grew best on lactose medium.

What can one conclude from this? Because the parameter studied (growth) is essentially a quantitative trait, and because the effect of lactose as a selective agent is not known (assuming lactose exercises any selective pressure at all), it is difficult to conclude anything. Galactosidase activity of control and phage-treated plants was assayed at 35°C and pH 7.2, conditions under which, it was said, the petunia galactosidases were inhibited whereas *E. coli* β-galactosidase was not, but the results revealed nothing. Error bars overlapped in all cases and there was no correlation between enzyme activity and growth on lactose. In fact, Hess himself refrained from drawing unequivocal conclusions from these data.

The next set of experiments with transducing phages involved the *gal* operon (Hess 1979). This was just a repeat of the work published earlier

(Hess 1978), except that differences between controls and phage-treated individuals were more marked. Again, growth in the presence of (this time) galactose was the parameter used to infer a genetic mechanism to explain differences between controls and experimental plants. No molecular evidence was presented regarding the persistence and expression of *E. coli* genes in the progeny of treated plants. A follow-up study (Hess and Dressler 1984) attempted to differentiate between normal plant enzymes that break down galactose and *bona fide E. coli* transferase (the product of the T gene in the *gal* operon). Unfortunately, these assays were based again on temperature and pH optimums without any published efforts to run much more specific identification experiments or even Southern blots to ensure that the gene was present. The year was 1984 and probes of all possible sorts were available to do these routine tests. What is more, this paper suggested induction of enzyme activity by galactose itself. Did this mean that the *gal* operon is inducible in plants? Could plants have the equivalent of a *gal* repressor gene whose product would recognize the *E. coli gal* operator? No explanation was offered. Once again, outdated biochemical techniques were used and only very circumstantial evidence (mostly growth on an unusable sugar) was provided. This type of research has since been abandoned in Dieter Hess's laboratory.[5]

From Russia, with Skepticism

Articles describing unusual results are not the sole apanage of the West. Two from the former USSR deserve mention. In the first (Turbin et al. 1975), a Russian/Byelorussian team set out to study the transformation of barley with homologous (not bacterial) DNA, using the *waxy* trait as a marker. The *waxy* mutation is recessive and is characterized by pollen grains containing the starch amylopectin but no amylose, contrary to the wild type. Amylose can be detected easily in pollen grains simply by staining them with a solution of iodine in potassium iodide. Wild-type pollen stains very dark blue, whereas *waxy* pollen stains reddish brown. The difference in staining is unmistakable and very specific. What is more, the *waxy* mutation is extremely stable and thus reverts at very low frequency.

DNA from wild-type plants was isolated and injected into barley grains, early during embryogenesis. The rationale was that injected exogenous DNA, carrying the dominant *Wx* (nonwaxy) gene, might become incorporated into the developing embryo, be expressed, and give plants forming easily detectable wild-type pollen. Because the donor DNA was also from barley, correct transcription and translation would not be a problem. Pollen grains isolated from the F2 generation of the injected plants (all plants were pooled) were stained with I_2/KI and microscopically examined. *E. coli* DNA (which of course does not contain any *Wx* gene) was used as a control. Interestingly, injection of water or *E. coli* DNA did increase reversion frequency about 10-fold over the spontaneous rate, and so did injection of highly deproteinized, highly polymerized barley DNA or low-molecular-mass, slightly deproteinized barley DNA. The authors concluded there was no effect at this point. However, injection of high-molecular-mass, slightly deproteinized barley DNA increased the frequency of wild-type pollen by a factor of 230. Since all treated plants had been pooled to score for the presence of wild-type pollen, it could not be determined whether one or a few single plants contained wild-type pollen only, or if all plants just contained a small amount of wild-type pollen each. Upon examination of individual plants, it turned out that one single individual contained 99.5 percent wild-type pollen and 0.5 percent mutant pollen. The authors declared this plant transformed but could not account for the presence of a small percentage of mutant pollen. Indeed, 0.5 percent mutant pollen would be about 1,000-fold the rate of spontaneous reversion. They also concluded that this plant could not have been a wild-type contaminant, in which case 100 percent of the pollen would have stained blue-black.

One question the authors could not address was that of the number of cells in the embryo that could have taken up DNA. Indeed, one complication is that more than a single cell gives rise to all the pollen grains seen in the mature plant, and, to observe 100 percent transformed pollen, one would have to stably transform 100 percent of the pollen precursor cells. The strangest aspect of this study, however, was that only "dirty," slightly deproteinized DNA worked. This is at variance with what we know today, which is that dirty DNA does not work well in transformation experiments. But how dirty was their DNA? They did not say. It would have been so easy to quantify the amount of protein, RNA, and so forth in their DNA preparations, but they did not do it. It is thus impossible to repeat these experiments under standardized conditions. The reason that only DNA complexed with proteins worked in transformation was attributed to a protection from DNase attack. This is somewhat plausible, but what was the effect of the "dirt" alone, not complexed with DNA? It is impossible to say.

Certainly, the *waxy* system had some potential, because only brute observation is required (they looked at over 4 million pollen grains), but there was no follow-up to these experiments. On the other hand, no selective pressure, so dear to geneticists, can be exercised in this system, making it impossible to enrich (at the expense of the background noise) the population of test organisms for sought-after genetic events.

In a subsequent paper (Soyfer and Titov 1981), the Russian portion of the team mentioned previously, used seed incubation with DNA (à la Ledoux) to determine whether characters for genetic necrosis and awnness (presence of spikelets on the inflorescence) could be imparted by exogenous DNA in wheat. They saw nothing in terms of necrosis but made an unsettling observation with awnness: Both the DNA extracted from the awned variety and that extracted from the awnless variety caused the appearance of awns on plants originating from DNA-treated seeds! In other words, the control and the test DNAs had the same effect. What is more, this modification was inherited at least till the second generation (they did not examine more advanced generations). Thus, DNA that was not supposed to introduce any genetic alterations into wheat plants did so and did it in a strange manner, one that does not make any sense at all.

Indeed, when an awnless plant made awned after treatment with DNA extracted from an awnless plant was selfed, segregation was observed. This would be expected if this plant had been a heterozygote (although the gene for awnness was never supposed to be present). However, only eight F1 plants were obtained, making it impossible to say if segregation was Mendelian. What was baffling, however, was that a selfed F1 awnless plant segregated awned and awnless individuals in F2! As the awned character is dominant, these results cannot be classically understood. The authors concluded that something other than simple genetic factors was at play. This game of allelic now-you-see-it-now-you-don't is somewhat reminiscent of the observations made with *Arabidopsis*, except that there, it was the mutant allele that disappeared and then reappeared, rather than the dominant character as here.

Conclusions

Transformation experiments conducted throughout the 1970s were all irreproducible, uninterpretable, or both. Why? We saw in chapter 1 that an intermediate-density DNA peak had been observed independently by several researchers. There, it was the *interpretation* of these results that had been shown to be incorrect, through the use of fine molecular techniques. The greater difficulties encountered here in chapter 2 have to do with the fact that *living systems*, not just molecules, were the subject of observation. Indeed, what were these early researchers looking at? Unlike DNA, which can be characterized very precisely through a number of variables such as density, molecular mass, strandedness, circularity or linearity, and sequence, living organisms cannot be easily encapsulated. The only model geneticists have through which they can interpret the behavior of a gene in the progeny of an individual is Mendel's, which is not without limitations.

The greatness and limits of Mendelian analysis can be summarized as follows. It is *because* the laws of Mendel are so general and their predictive quality is so high that it is possible to question the interpretation of results that violate these laws. However, violation of these laws does not necessarily mean that a trait is not heritable at all. All this says is that the observed phenotypes are probably not under the control of a Mendelian gene—that is, a factor with totally predictable behavior in controlled crosses. This of course is a tautology, but it is a highly successful and heuristic one. Does this mean that Mendel's laws apply only to certain categories of genes? Certainly not; but the genius of Mendel was to choose for his experiments traits that contrasted maximally and were not influenced by the environment. His pea plants were either tall or short and uniformly so, and his seeds were yellow or green, not something in between. In other words, he decided to work on objects that fell into sharply defined phenotypic categories. This is what allowed him to derive his laws.

Are all phenotypes sharply categorizable? Of course not, and this is where difficulties crop up. Many traits are said to be quantitative—that is, they will vary from one extreme to another without simple categorization. In these cases, a *number* of genes are involved in the determination of the phenotype. Further, many genes are semidominant—that is, when present in only one copy in an individual, they will not impart to this individual a fully dominant phenotype; these genes are thus *additive*. The segregation patterns of multiple and additive genes do remain Mendelian, however. It is just that the situation can become so complicated that a simple model can no longer be built. Finally, some genes are *epistatic* to others, meaning that they control the expression of other genes. These regulatory genes complicate things even further. If all those gene interactions are taken into account, plus the fact that environmental factors can influence gene expression, it becomes clear that significant problems could be encountered by choosing a complex (as seen with 20/20 hindsight) experimental system.

This is what happened in the experiments with petunia flower color. Pigment synthesis in flowers is complicated; dozens of genes can be involved and, as we have seen, their activity can be strongly influenced by changes in the environment, as well as by viral infection. Surely, unraveling all these intricate pigment biosynthetic pathways and their genetic control can be exciting, but this was not the objective of the work of Hess. He wanted to demonstrate an entirely new principle: *that externally supplied DNA could modify the genotype of an individual in a heritable manner.* His choice of material was simply not suitable: The complex petunia flower pigmentation system could not possibly provide any answers to this question. Likewise, the choice of growth rate on lactose and galactose as an indicator that plants have been heritably transformed by foreign DNA was unsuitable, as it is inconceivable that this trait, in green plants, can be understood in terms of the effects of a single gene.

The Ledoux team was after the same goal, the DNA-mediated transformation of plants. Ledoux's thought process was different from that of the other geneticists whose work is described in this chapter. He was misled by the earlier, erroneous physicochemical evidence that seemed to imply the presence of large amounts of bacterial DNA integrated within and replicating with plant host DNA. This led him to expect that transformation events would be frequent. Moreover, and he could not be blamed for believing this, he thought that bacterial genes should be able to complement mutations in plants. We now know that this is not the case; prokaryotic genes must be engineered with plant promoter regions to be expressed in a plant environment. This of course was not known at the time.

Thus, biological effects of bacterial DNA in *Arabidopsis* were seen (although by only one group), but these effects defied Mendelian analysis. This should have been a red flag, as the thiamin pathway in plants is considerably less complicated than pigment pathways. At least, thiaminless mutants of the type used by Ledoux did segregate in a Mendelian way and were not influenced by environmental factors. The problem is that the "transformants" did not segregate properly and were influenced by external factors such as temperature. As nobody else could reproduce these effects, one can only conclude that indeed they were the result of contamination by other seed stocks or, alternatively, by pollen from other mutants or the wild type. In retrospect, these experiments could not yield positive results other than by contamination, an aspecific mutagenic effect of the donor DNA, or other uncontrolled factors. The donor genes, we now know, were just not of the right kind to correct a genetic defect in a plant.

In the two examples discussed here, the authors' failure to require their "transformants" to adhere strictly to Mendelian ratios led them astray. For example, meiotic drive (a rare occurrence) was repeatedly invoked to explain away unusual ratios in crosses. Further, whenever unexpected lethality occurred, it was always preferentially among the categories that *should* have been present and were not. Clearly, even in the age of molecular biology, old-fashioned Mendelian analysis must remain the powerful and crucial tool that it always was.

We now know that these early transformation experiments could not have worked. In the case of bacterial donor genes, promoters were of the wrong kind and transformation frequencies were much too high to assume accidental integration of the prokaryotic transgenes next to a plant promoter. In the case of petunia flower color, the donor DNA was of homologous origin and thus potentially expressible. There, it is gene dosage that was much too low. Further, there is still no evidence that the DNA uptake techniques used by Hess and Ledoux actually work. Gene cloning and efficient DNA transfer methods had to be developed first, to demonstrate DNA-mediated transformation of plants.

A positive aspect of this enormous controversy, however, was to draw attention to an undeveloped area of science: the genetic transformation of plants. In that sense, all the publicity and the controversy were not for naught. In the end, however, scientific knowledge has to become paradigmatic and consensual. This ultimately did take place in the field of naked DNA–mediated transformation of plants, in spite of its very difficult beginnings. This success story will be told in chapter 4. And indeed, petunia flower color *has* since been manipulated by direct gene transfer.

CHAPTER 3

The Crown Gall Breakthrough

As the events described in chapters 1 and 2 were unfolding, another avenue to introduce foreign DNA into plants was being discovered: the now classical *Agrobacterium*-mediated gene transfer process. Thus, the first unequivocal demonstration that DNA can be transferred to plant cells came in fact from the study of a seemingly mundane plant pathogen, the soil bacterium *Agrobacterium tumefaciens*. Although fundamentally different from the notion that plants could conceivably be genetically transformed by uptake of isolated DNA, this new gene transfer process gave confidence that DNA-mediated transformation might one day become a reality as well. However, the study of *Agrobacterium* and crown gall was not originally intended to lead to the genetic transformation of plants; this became a goal only after the nature of crown gall disease had been partially understood.

It had been known since the turn of the century that *A. tumefaciens* was the causative agent of crown gall, a neoplastic disease of dicotyledonous plants, characterized by undifferentiated tumor formation, usually at the site of a wound (fig. 3.1). The work of Armin Braun of Rockefeller University in the 1940s clearly demonstrated that tumor tissue could be propagated *in vitro* under axenic conditions, meaning that once tumorigenesis has been incited, the causative agents, the bacteria, are no longer needed. Thus, *A. tumefaciens* is necessary for tumor initiation but not for tumor maintenance and proliferation. These cells are thus permanently "transformed." Further, crown gall tumors can be propagated on defined culture medium without phytohormones, auxins, and cytokinins, conditions under which normal plant tissue (except habituated tissue, a very rare occurrence) will simply not survive. Also, tumors cannot be forced to redifferentiate into morphologically normal plants through the action of externally added plant hormones. Somehow, they are "stuck" in a permanent undifferentiated or semidifferentiated state.

FIGURE 3.1.
Crown gall tumors on carrot slices.

Braun first posited the existence of a "tumor-inducing principle" (TIP) that must pass from the bacterium to the plant cells and bring about these permanent changes. Finally, crown gall tissues synthesize unusual amino acids, called opines, which are not normally found in healthy plant tissues, as shown by the pioneering work of Georges Morel in France in the 1950s. Crown gall cells are very peculiar plant cells indeed.

It took over three decades to solve the crown gall riddle and answer the following four questions: (1) What is Braun's tumor-inducing principle, which stably "transforms" normal plant cells into crown gall tumor cells? (2) How do crown gall cells multiply without phytohormones? (3) What is the role of wounding? and (4) Why do crown gall cells synthesize unusual amino acids (opines) such as octopine and nopaline, for example. Answers to these questions made *Agrobacterium*-mediated gene transfer to plants a reality and, in fact, there is a single answer to all four questions: the Ti plasmid. A great paradigm had finally reached the realm of plant biology. But the birth of this concept came about with labor and pain, as plant biologists had by now come to expect.

Most reviews of the beginning years of *Agrobacterium* transformation have presented a linear story, where one discovery follows another and inspires the next. However, things did not happen that way at all, and certainly not in the beginning. Starting about 1975 or 1976, very stiff competition between labs prevented publication of any baroque findings and kept everybody alert to gross misinterpretations. The volume of literature describing the advances in our understanding of the crown gall phenomenon is substantial and well reviewed by others. Full citation of all publications is outside the scope and intent of our story here. Instead, I will cite only a few key, seminal articles in the references. In the citing of articles, I have attempted to keep a balance between the contributions of various labs. Sometimes, it was impossible to state who confirmed whose work, given natural delays in publication and private communications made at meetings, over the phone, through students, visitors, and postdoctoral fellows. This was a competitive field indeed.

Confusion

The search for the TIP was not an easy one. Because the crown gall phenotype was so stable, and because *Agrobacterium* could be eliminated once it had triggered tumor formation, it was initially thought by most that the plant cell genome itself had been modified in some way. But how? Could it be by mutation? This would not easily explain the pleiotropic effects seen: opine synthesis, hormone independence, and permanent dedifferentiation. Granted, dedifferentiation and hormone independence could be correlated, because crown gall cells produced their own auxin and cytokinin, precisely in ratio and amounts that kept these cells from differentiating again.

But then where did the opine genes come from, and what was it that allowed these cells to produce their own hormones when normal cells in culture did not do that? One would have to assume multiple mutations to explain all these phenomena, and it was hard to imagine how bacterial cells would have the ability to mutagenize always the same plant genes. Further, there existed several *Agrobacterium* strains, some not even pathogenic (i.e., not causing crown gall). What was the difference between pathogenic and nonpathogenic strains? Finally, virulent *Agrobacterium* strains were able to convert nonvirulent strains to a virulent state when multiplying together in a plant. Surely, in the latter case, genetic information was passed from one strain to another.

Another intriguing observation was that *Agrobacterium* is able to utilize its own opines as carbon and nitrogen sources. That is, the bacteria catabolize what crown gall cells produce. Furthermore, an *Agrobacterium* strain able to catabolize octopine (this strain is then not able to grow on nopaline) produces crown gall tumors synthesizing the very same amino acid, and the same holds true for nopaline. What was it that a bacterial strain had that allowed it to utilize the same opine that it caused the tumor cells to produce? These questions had been posed by 1970 by Morel's group at the Institut National de la Recherche Agronomique in Versailles, France, but there were as yet no universally accepted answers.

Because live *Agrobacterium* was not needed to maintain the tumorous phenotype, it is likely that many a researcher was reminded of Griffith's 1928 experiment in which he demonstrated that pathogenic, heat-killed *Streptococcus pneumoniae* could transfer a "virulence principle" to a nonpathogenic strain and transform it into a pathogenic version. Everybody who has taken a general genetics course knows about this experiment and will remember that the "virulence principle" is indeed DNA. Was the same thing happening in the case of crown gall? But this would have meant that a prokaryote (*Agrobacterium*) was able to transfer genetic information (DNA or RNA) to a eukaryote (the plant)!

This was without precedent in biology: Except with bacteriophages and other viruses, such transfer of genetic information had never been seen. This type of transfer, now thought to have played a significant role in evolution, is called horizontal (or lateral), in contrast to the normal vertical passage of genetic information from parents to progeny either by mitosis or by sexual reproduction. We also know now that horizontal transfer of DNA in the form of plasmids is a common occurrence among bacteria in nature (see appendix 4). However, the idea of horizontal gene transfer across kingdoms—bacteria to plants—was viewed with distrust at the time and was considered by many too heterodox to be taken seriously.

Yet we will see that this is exactly what is happening; *Agrobacterium* is able to donate some of its DNA to plant cells through a mechanism equivalent to mating among bacteria, except that mating here is taking place between a bacterium and a plant cell. This remarkable transfer of DNA between totally unrelated organisms is so far the only one of its kind known in nature. But first let us go over some of the early claims made about the TIP.

For all the reasons just summarized, crown gall formation seemed genetic in origin, thus involving DNA or RNA. But what was its source? And how did it cause all these effects? Between 1970 and 1976, several hypotheses were formulated and tested, and all were rejected. For example, a group had detected the presence of phage PS8, which infects certain strains of *Agrobacterium*, in crown gall tumors. Was this the TIP? Well, some other research groups claimed that indeed, this phage was tumorigenic and that its DNA could be detected in tumor cells by nucleic acid hybridization. These results were quickly proven wrong by others; this phage had nothing to do with crown gall formation. Was *Agrobacterium* DNA itself the TIP? Here again, claims were made that purified total *Agrobacterium* DNA could indeed incite crown gall and could even be detected in the crown gall cell genome by hybridization. Again, these claims were tested and overturned.

If phage and *Agrobacterium* DNA were not responsible, maybe *Agrobacterium* RNA was. One team published that, yes, small RNA molecules from *Agrobacterium* were tumorigenic and could be reverse-transcribed into DNA in the host. This claim was never sustained either. In fact, there is to date no evidence that prokaryotes can reverse-transcribe RNA. The reader must now wonder (as did many investigators in that period of our history) whether the *Agrobacterium* story was just a reiteration of chapters 1 and 2. Indeed, there are many similarities between the saga of DNA uptake in plants, and the vagaries of molecular research on crown gall at its beginning. The big difference, however, is the number of research groups involved. Many more people were interested in crown gall than in pure DNA uptake by plants for a simple reason: Crown gall was a genuine and mysterious natural phenomenon to be understood, whereas DNA uptake and expression by plants was a wild goose chase. Thus, in the case of crown gall, wild claims were rapidly tested by others and rejected if wrong.[1] Many articles proving and then disproving the role of *Agrobacterium* DNA and RNA were published in the span of two or three years. They are too numerous to cite here but can be found in Lippincott and Lippincott (1975). Fortunately, this state of affairs was to change radically, from a modest start, in 1974.

Logic at Last: Discovery of Plasmids in *Agrobacterium*

That year, the Schell/Van Montagu laboratories, both at the University of Ghent, Belgium, had come up with an exciting discovery, that of large plasmids in virulent *Agrobacterium* (Zaenen et al. 1974). This was not a trivial piece of information. Plasmids had been isolated from *E. coli* and some other gram-negative bacteria several years before. However, these were fairly small DNA molecules, measuring perhaps 20,000 base pairs (20 kbp) maximum. To isolate these molecules, gentle lysis of the bacterial cells was needed, lest the plasmids be sheared and lost, but *Agrobacterium* did not respond well to these gentle methods. Patient optimization of bacterial cell lysis by the Schell/Van Montagu group paid off, and huge plasmids, the largest so far, were detected, with a size of about 200 kbp. What did these plasmids do in *Agrobacterium*?

The genus *Agrobacterium* contains several species, of which three are pathogenic (*A. tumefaciens, A. rhizogenes,* and *A. rubi*) and one is not (*A. radiobacter*). The team studied strains from all four species and, sure enough, large plasmids were found only in pathogenic strains. They immediately concluded that, even though their experiments did not demonstrate that these plasmids were the TIP, there was a one-to-one correlation between ability to induce a pathogenic response in plants and presence of a large plasmid in the inciting bacteria.

In science, correlation is not proof. That is, it could still be that the existence of large plasmids in virulent *Agrobacterium* was coincidental and that other factors were responsible for plant tumorigenesis. This is where the power of bacterial genetics came into play. It was known that growing bacteria at a barely sublethal temperature caused frequent loss of plasmids by lack of replication. Bacterial strains so treated are called "cured" of their plasmid. It was known that incubating virulent *A. tumefaciens* at 37°C (*Agrobacterium* grows best at 29°C) caused loss of virulence. The Schell/Van Montagu group thus cured virulent strain C58 and, indeed, loss of virulence was associated with loss of plasmid (Van Larebeke et al. 1974). But again this was only negative correlation, although the evidence was becoming really strong. The next logical step was to take a cured, avirulent strain and demonstrate *gain* of virulence due to *gain* of plasmid. This was done by infecting plants with a mixture of virulent (plasmid-containing) bacteria and avirulent (devoid of plasmid) bacteria, and indeed, the originally avirulent cells became pathogenic after acquisition of the large plasmid (Van Larabeke et al. 1975). What had happened here was plasmid transfer by conjugation between the two types of bacteria inside the plant wound (appendix 4). It was later shown that plasmid transfer by conjugation between *Agrobacterium* cells could also be achieved completely *in vitro*, without the need for a plant host (Hooykaas et al. 1977).

By 1975, there was no doubt that the large *Agrobacterium* plasmid played a critical role in crown gall formation. However, all the results had so far come from the Ghent team (with help from Schilperoort, University of Leiden, The Netherlands) and had not been independently confirmed. This was not to last: A U.S. team from the University of Washington in Seattle, with Eugene Nester, Milton Gordon, and Mary-Dell Chilton at the helm, appeared on the scene and forced the Ghent and Leiden people into a competitive race that in just a few more years would lead to a crowning achievement, the production of transgenic plants. But there was still much basic work to do. The Seattle team did indeed confirm the role of plasmids in tumorigenesis in 1975 (Watson et al. 1975). Also, the ability of *A. tumefaciens* to catabolize opines and to determine their synthesis was linked to the large plasmid, which soon became known as pTi or Ti plasmid (p stands for plasmid and Ti for tumor inducing) (Bomhoff et al. 1976). The Ti plasmid started to look very much like the TIP, but how exactly did it all work? This was still very unclear, but first speculations on the use of the Ti plasmid for plant genetic engineering had already been formulated (Schell 1975). Ironically, serendipity was also at work here; it was discovered later that avirulent strains of *Agrobacterium* could carry plasmids even larger than pTi (up to 300 kbp) (Merlo and Nester 1977)! These had not been detected earlier for technical reasons. Fortunately, these plasmids play no role in tumorigenesis. Still, this was a close call.

Racing Ahead: The Elucidation of the Crown Gall Problem

By 1976, three competing groups were dominating the field of crown gall research. It was also around that time that the Schell/Van Montagu synergistic collaboration really took off. Also, thanks in part to the notoriety gained through his association with the Ghent group, Robb Schilperoort was able to build a strong research team in Leiden very quickly. As for the Seattle group, its strength came in part from the merging of two labs, that of Nester, who was a bacterial geneticist and that of Gordon, who was a plant biochemist. Mary-Dell Chilton had a courtesy appointment in the Department of Microbiology and Immunology and was well versed in genetic and molecular techniques. Interestingly, the initial goal of the Seattle group, in 1971, was to reevaluate the claims of DNA integration in plants as described in chapter 1, as well as to test the validity of all the early results of DNA and RNA hybridization experiments with crown gall DNA.

The T-DNA from pTi

The first definitive demonstration that pTi was indeed the TIP was achieved in Seattle. The term *TIP* immediately became obsolete, as this was now a known entity. Here, also, there was at first some controversy regarding the presence of pTi sequences in the crown gall genome. Matthysse and Stump (1976) had reported the presence of a significant amount of pTi sequences in crown gall DNA, but this result could not be repeated in Seattle. It was concluded that if any pTi DNA was present in crown gall, it had to be less than the whole Ti plasmid itself. With hindsight, the reason for so many flawed DNA experiments was that crown gall tumor cells were fed a rich diet of sucrose, so they produced enormous amounts of polysaccharides, which copurified with tumor DNA and caused all kinds of trouble in DNA hybridization experiments. During a sabbatical stay in the Schilperoort laboratory in 1974, Gene Nester had already demonstrated totally aspecific hybridization with impure crown gall DNA.

The University of Washington team thus decided to check the solution hybridization pattern of restriction endonuclease fragments of pTi with highly purified crown gall DNA. Now, the Ti plasmid is so big that most restriction endonucleases (and certainly the ones available at the time) will cleave it into many fragments, some of similar sizes, making it difficult to separate them by agarose gel electrophoresis. The *Sma* I restriction enzyme was selected because it cleaved pTi into only (!) 19 bands, some of them multiplets. Radioactive pTi restriction bands were thus gel-fractionated, eluted from the gel, and used to run P_0t curves with crown gall and control DNAs. One band, 3b, showed dramatic acceleration of reassociation in the presence of crown gall but not control DNA (fig. 3.2) (Chilton et al. 1977). That was it! A small portion of the Ti plasmid was actually present in the genome of the tumor cells, and it could be responsible for all the phenotypic changes seen in crown gall tissue.

Incredibly, the unthinkable had been demonstrated: *Agrobacterium* was able to cross an interkingdom barrier and deliver genes to a eukaryote. This article (Chilton et al. 1977) quickly became a "citation classic." But in what state was this T-DNA (for transferred DNA, as it became known) present in crown gall cells? Was it free, like a plasmid? Was it integrated within the plant DNA? Where was it located? In the nucleus, in cytoplasmic organelles? Was it transcribed? And, finally, what functions was it coding for? Could it indeed determine the synthesis of opines and could it possibly code for the synthesis of phytohormones found in the tumor cells? The next several years saw all these questions answered, and for once there was agreement among the competing groups as to what the answers were!

Transcription of the T-DNA in crown gall cells was demonstrated by Southern hybridization of radioactive RNA from crown gall cells with Ti plasmid digested with *Sma* I, as mentioned. Band 3 lit up, indicating that tumor cells did indeed synthesize RNA molecules complementary to T-DNA (Drummond et al. 1977). It is now known that the T-DNA of some of the Ti plasmids codes for up to 13 different mRNAs, most of them not known even today to be involved in tumorigenesis. Additionally, the Seattle group

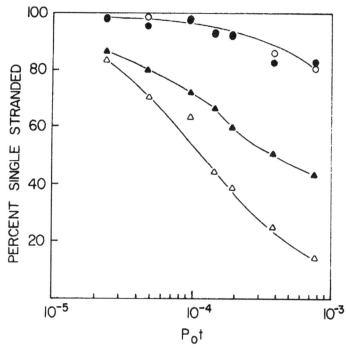

FIGURE 3.2.
Renaturation kinetics of a portion of pTi (Sma I band 3b) in the presence of control and crown gall DNA. *Open circles*: Reaction driven by salmon DNA. *Closed circles*: Reaction driven by normal tobacco DNA. *Closed triangles*: Reaction driven by tobacco crown gall DNA. *Open triangles*: Model reconstruction experiment in which the reaction is driven by salmon sperm DNA supplemented with 10 copies of pTi DNA Sma I band 3b. It can be seen that salmon and normal tobacco DNAs do not accelerate the reaction, whereas that from crown gall does. The reconstruction experiment shows that pTi Sma I band 3b is present at less than 10 copies per crown gall genome. (Adapted from Chilton, M.-D., M. H. Drummond, D. J. Merlo, et al. 1977. Stable incorporation of plasmid DNA into higher plant cells: The molecular basis of crown gall tumorigenesis. *Cell* 11:263–271.)

demonstrated that a particular crown gall line, BT-37, was able to produce a fertile shoot when grafted to a healthy tobacco plant. Analysis of the normal, F1 progeny of this plant revealed complete absence of T-DNA sequences, demonstrating that meiosis could cause elimination of the foreign DNA. These experiments also established that continued presence of the T-DNA was absolutely necessary for the maintenance of the tumorous state (Yang et al. 1980). It had thus been clearly shown that gain of T-DNA led to a tumorous condition, whereas its loss caused plant tissue to revert to a normal phenotype.

The question of the physical location of the T-DNA was tackled next. Plant cells contain three different genomes, segregated into three different organelles, the nucleus, the chloroplasts, and the mitochondria. There was no reason to suppose *a priori* that the T-DNA was in only one, two, or all three types of organelles. Thus, cellular organelles were separated by differential centrifugation and purified; their DNA was isolated, digested with restriction endonucleases, and electrophoresed in agarose gels. Fragments were then Southern blotted and hybridized with a radioactive DNA fragment from the T-DNA. Only the nuclear DNA lit up, indicating that the T-DNA was in the nucleus and not in mitochondria nor chloroplasts (Chilton et al. 1980; Willmitzer et al. 1980).

As mentioned, the T-DNA, now known to be present in nuclei only, could conceivably behave like a bacterial or yeast nuclear plasmid; that is, it could remain independent from the chromosomal DNA and also replicate independently. The year was 1980, and by then, the T-DNA had been mapped with restriction endonucleases and restriction fragments had been cloned. The status of the T-DNA in crown gall cells was investigated as follows: Crown gall DNA was digested with restriction endonucleases into about 10^6 fragments. These were cloned in bacteriophage λ, and these clones probed with T-DNA fragments in Southern blots. The λ clones contained T-DNA fragments that were too long, as if these portions of the T-DNA were perhaps linked to plant DNA. Was this the case? Yes, when the extra (non-T-DNA) DNA was cut out of the λ clones and used to probe normal plant DNA, it lit it up. This meant that the λ clones contained, in addition to T-DNA, plant DNA, *to which the T-DNA was covalently attached* (Yadav et al. 1980; Thomashow et al. 1980). It was also later demonstrated that the T-DNA was present in a nucleosome structure: that is, DNA wrapped around histone cores, like any other eukaryotic nuclear DNA.

It is perhaps time to summarize what we have learned. Virulent *A. tumefaciens* harbors a large Ti plasmid and somehow (in an as yet undefined way) transfers a portion of it (14 kbp in the case of octopine strains, or 24 kbp in the case of nopaline strains), called the T-DNA, to the nucleus of plant cells, where this T-DNA becomes integrated, covalently joined to the plant DNA. In effect, this piece of bacterial DNA becomes an integral part of the plant genome. Moreover, the T-DNA is transcribed into mRNA. This is exactly what the research described in chapters 1 and 2 tried to achieve by using purified exogenous DNA, which had not yet been successful. *It turned out that Agrobacterium is a natural plant genetic engineer.*

Of course, *Agrobacterium* has evolved to engineer plants for its own purposes, not for human purposes. But could humans fool *Agrobacterium* by disarming it, neutralizing its tumor-causing T-DNA, and replacing it with useful genes? Before answering that question, it is necessary to understand where exactly in the T-DNA these tumor-causing (*onc*, for oncogenic) genes are and what they do.

The Function of T-DNA Genes

Ti plasmids isolated from different *Agrobacterium* strains do not necessarily share many common sequences; their homology is limited. However, high sequence homology is found in two regions of these plasmids: the T-DNA, and what was later named the *vir* region, a long stretch of Ti plasmid totally distinct from the T-DNA (see later). Both the T-DNA and the *vir* region are crucial for tumor formation. The functions of the genes located in the T-DNA were studied by transposon mutagenesis. In these experiments, transposons were introduced by conjugation into *Agrobacterium* cells where they "jumped" (see appendix 4) and inserted themselves randomly into the chromosome and also into pTi. When transposons insert themselves into the middle of a gene, they interrupt and inactivate it. Thus, when transposons transpose into T-DNA genes, they are expected to interfere with tumorigenesis, and *Agrobacterium* harboring such transposon "hits" might not generate crown gall. When *Agrobacterium* strains containing transposons at different places in the T-DNA were used to infect plants, three types of tumors were found: those that no longer produced opines (octopine or nopaline) but had otherwise a tumor phenotype, those that developed shoots at the wounding site, and those that developed roots at that site. The conclusions were obvious: First, the gene coding for opine synthesis is part of the T-DNA, and this explains why crown gall tumors produce opines. What had happened in the mutagenized pTi was that a transposon had interrupted this gene, which ceased to function, and hence no opine production was detected. This finally explained why crown gall tumors are able to synthesize these unusual amino acids: Their genes are carried by the T-DNA. Second, depending on where the transposon had hit, "shooty" or "rooty" tumors were obtained. But in no case was there any effect on T-DNA transfer. Genes needed for this transfer were *not* a part of T-DNA.

What was the explanation for the shooty and rooty tumors? It had been known for a long time that normal plant cells in tissue culture need a certain balance between the two main plant hormones, auxins and cytokinins, to grow as undifferentiated masses—in a way, imitating crown gall. Also, it had been observed that increasing the ratio of auxins to cytokinins led to root formation in culture, whereas increasing the ratio of cytokinins to auxins led to the formation of shoots. Thus, the appearance of rooty and shooty tumors very strongly suggested that T-DNA carried genes that code for the production of auxins *and* cytokinins. When an auxin gene was interrupted by transposons, auxin production would stop but that of cytokinin would continue. Result: High cytokinins would trigger the appearance of shoots. Similarly, if indeed the T-DNA also coded for cytokinin production, if a cytokinin-producing gene was inactivated by transposon insertion, this would lead to production of auxins only, and the tumor would have a rooty appearance (fig. 3.3).

A lot of work then followed to identify and characterize these *onc* genes (Ooms et al. 1981; Garfinkel et al. 1981; Leemans et al. 1982), and the hypotheses regarding their functions were verified; it turned out that a single gene, gene 4 (also called *roi* or *tmr* or *ipt*Z), determined the production of isopentenyladenosine-5'-phosphate, which is a cytokinin, while two genes, called genes 1 and 2 (or *shi* 1 and 2, or *tms* 1 and 2, or *iaa*M and *iaa*H), were necessary for the production of indole acetic acid (IAA), the natural plant auxin. Thus, inactivation by transposons of the cytokinin-producing *ipt*Z gene produced rooty tumors, whereas inactivation of *iaa*M and *iaa*H produced shooty tumors. This all made complete sense.

The mystery of crown gall had been solved. *Agrobacterium* causes tumor formation in plants because its T-DNA disrupts the hormonal balance in plant tissues by forcing the cells to produce auxin and cytokinin in ratios leading to undifferentiated cell proliferation. Opines play no role in tumorigenesis proper, and the *nos* or *ocs* (for nopaline synthase or octopine synthase) genes are dispensable. These genes still play a role in nature, though. It is thought that the presence of an opine gene in the T-DNA and its expression in tumors is evolutionarily advantageous to *Agrobacterium*, because these bacteria can use opines as carbon and nitrogen sources. Thus, once tumor formation has been triggered, *Agrobacterium* will find itself in a nutrient-rich environment and proliferate abundantly. This phenomenon was termed genetic colonization. Further, certain opines (octopine for one) turned out to be inducers of conjugation of Ti plasmids. This explains why plasmid transfer occurs in the infected plants cells: The inducer is present! Figure 3.4 shows the general organization of genes in octopine and nopaline types of T-DNAs.

The basic nature of crown gall disease had thus been elucidated and confirmed independently by 1982. This year may be seen as a turning point in plant genetic engineering to the extent that a plant transformation vector, the T-DNA, had been discovered and its physiological effect, tumorigenesis, understood. This year also marked, I think, the beginning of the end of an intense period of fundamental discovery. Many questions remained to be

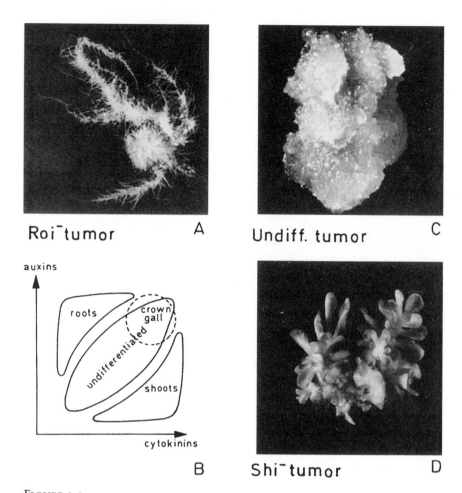

Roi⁻tumor   **A**

Undiff. tumor   **C**

auxins

roots   crown gall

undifferentiated

shoots

cytokinins

B   **Shi⁻tumor**   **D**

FIGURE 3.3.
Roi⁻ (rooty) tumor (**A**), crown gall (undifferentiated tumor) (**C**), and shi⁻ (shooty) tumor (**D**). The diagram (**B**) shows morphological changes induced by varying the auxin versus cytokinin ratios in normal plant tissues. These changes correspond to the characteristics seen in the tumor photographs. The crown gall domain is represented by an interrupted circle (From Gheysen, G., P. Dhaese, M. Van Montagu, and J. Schell. 1985. DNA flux across genetic barriers: The crown gall phenomenon. In B. Hohn and E. S. Dennis, eds. *Genetic Flux in Plants*, pp. 11–47, figure 3. Vienna: Springer-Verlag. Copyright Springer-Verlag. Reprinted by permission of the publisher.)

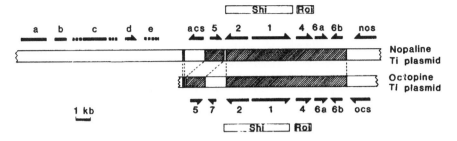

FIGURE 3.4.
Comparison between the T-DNA regions of octopine and nopaline Ti plasmids.
Homologous sequences are represented by hatched boxes. *Arrows* indicate direction
of transcription, and *jagged lines* define the T-DNA borders. acs, agrocinopine syn-
thase; nos, nopaline synthase; ocs, octopine synthase; Shi, shoot inhibition locus;
Roi, root inhibition locus. (From Caplan, A., L. Herrera-Estrella, D. Inzé, et al. 1983.
Introduction of genetic material into plant cells. *Science* 222:815–821, figure 2.
Copyright American Association for the Advancement of Science. Reprinted by permission of the publisher.)

solved, such as the T-DNA transfer mechanism, but later work could be
viewed as being of a more "linear", that is, more predictable nature. And
indeed, it only took one single year to produce the first transgenic plant tissues using the T-DNA as a foreign gene vehicle (see later section, Victory).

Transfer of T-DNA from Agrobacterium to Plant Cells

T-DNA in independently incited crown gall tumors is, with exceptions, the
same length. This means that there must exist a mechanism in *Agrobacterium* that recognizes the T-DNA in pTi, cuts it at the same place most of
the time, and transports it out of the bacterial environment, into the plant
nucleus. Does this mechanism measure the length of the T-DNA? The
answer here is no, because transposons integrated within the T-DNA can
make it several thousand base pairs longer, but they are found in the crown
gall genome, still physically present within the T-DNA. The signal to cut the
T-DNA at a given position must then not be length alone. Sequencing of
the T-DNA itself (Gielen et al. 1984) as well as sequencing of the plant
DNA–T-DNA junctions in crown gall showed that the T-DNA is bordered
by an imperfect 25-base-pair repeat sequence (Yadav et al. 1982). This
repeat sequence is recognized by Ti plasmid–encoded enzymes as cleavage
sites. Sometimes these enzymes will make mistakes and cut not exactly at
the T-DNA borders, in which case longer or shorter T-DNA is transferred to

the plant. Or T-DNA molecules could occasionally be broken in transit, leading to truncated copies.

What is the mechanism in *Agrobacterium* that recognizes the 25 base pairs present at either end of the T-DNA? Transposons inserted anywhere in the T-DNA do not prevent its transfer to the plant cell nucleus. This means that T-DNA genes are not involved in excision and transfer. Then which genes do that job? While doing transposon mutagenesis experiments, researchers noted that a region well outside the T-DNA, but common to all Ti plasmids, was responsible for T-DNA excision and transfer. They found that transposon insertions in this region completely blocked crown gall formation (Iyer et al. 1982). This region was called *vir*, for virulence, and it was even shown that the T-DNA and the *vir* region need not be present in the same plasmid for the T-DNA to be transferred to plant cells (Hoekema et al. 1983; DeFramond et al. 1983). This is because the *vir* genes act in *trans*, meaning that they code for enzymes able to diffuse through the bacterial cytoplasm and exercise their function where needed—that is, at the level of the T-DNA borders where cutting occurs, or at the membrane site where the conjugational apparatus is assembled.

Genetic and molecular biological studies subsequently showed that the *vir* region is complex: It is composed of seven operons (*vir* A through G), each consisting of one to eleven genes (some being simply open reading frames). Six of these operons are absolutely required for virulence, but *virF* is not (Stachel and Nester 1986). They act in concert with chromosomal *chv* genes to effect the excision and transfer of the T-DNA to plant cells. Remarkably, one of the gene products of the *vir* region (*virA*) is an environmental sensor that interacts with a class of molecules called methoxyphenols (of which acetosyringone is the most well-studied representative) and that are precursors for the synthesis of plant cell walls (Stachel et al. 1985). Thus, wounded and dividing plant cells produce and release these compounds in sufficient amounts to interact with the *Agrobacterium* surface. The product of the *virA* gene is an allosteric protein, present in *Agrobacterium's* cell membrane, that is in contact with the external environment. VirA protein is capable of autophosphorylation only after contact with acetosyringone or other methoxyphenols, found in the vicinity of wounded or dividing cells. In that phosphorylated state, the virA protein in turn phosphorylates the product of *virG*, which, after also undergoing an allosteric transition, becomes a transcriptional activator of all *vir* genes except *virA*, which is expressed constitutively. Thus, *virA* and *virG* regulate the transcription of all other *vir* genes, and this regulation depends on the presence of acetosyringone or other environmental triggers (Stachel and Zambryski 1986).

That is, once acetosyringone has activated the virA/virG protein duo, all *vir* B, C, D, and E genes are turned on and expressed.

What are the functions of these *vir* products whose expression is stimulated by the acetosyringone–virA–virG cascade? The *vir*B operon codes for several proteins that are analogous to those found in the conjugation tube of bacteria, further reinforcing the idea that T-DNA transfer from *Agrobacterium* can be viewed as the mating of a bacterium with a plant cell. The putative nature of other *vir* products has been inferred by sequence homology with known proteins and then confirmed experimentally. VirE2 is a single-stranded DNA binding protein, and virD2 is an endonuclease. The analogy with bacterial conjugation goes even further: We now know that virD2 makes a single-stranded nick right at the border of T-DNA, which is followed by the progressive release of a single strand from the Ti plasmid until the other border is met, at which point the single-stranded T-DNA is released (Stachel et al. 1986). VirD2 is helped in the process by a topoisomerase (an enzyme able to relieve stress induced in DNA by the separation of the two strands). The virE2 protein then possibly "coats" the single-stranded T-DNA and accompanies it in its transit through the conjugation tube. This is how conjugation between two bacterial cells functions! It also turns out that virD2 becomes covalently bound to one of the ends of the single-stranded T-DNA (now called the T-strand) (Ward and Barnes 1988).

Interestingly, T-DNA processing is one of the few examples where controversy arose in the history of advanced crown gall research. It was initially thought that the T-DNA was removed from pTi as a *double-stranded* molecule that then circularized, plasmid-like, before transfer to plant cells (Koukolikova-Nicola et al. 1985). Results from the same Ghent lab that published the preceding data (Stachel et al. 1986) quickly showed, however, that this was not the way it all worked, and confirmation that the T-DNA was converted into a linear single T-strand before transfer came rapidly from the Seattle lab (Albright et al. 1987). It is now generally agreed that T-strand formation is the primary mechanism through which the T-DNA is transferred to plant cells, but the significance of circular T-DNA, found in *Agrobacterium* too, is unclear. These molecules could be a byproduct of T-strand formation, or they may play an as yet unknown role (Timmerman et al. 1988).

Another function of the virD2 protein was discovered several years later (Herrera-Estrella et al. 1990). VirD2 contains nuclear targeting signals—that is, short amino acid sequences known to be present in proteins, which recognize the nuclear membrane and allow these proteins to penetrate into the nucleus. The picture that we now have is that the T-strand, covalently bound to virD2, transits from the *Agrobacterium* cytoplasm, through a conjugation tube, to reach the plant cell cytoplasm, where it complexes with virE2 single-strand binding protein. VirE2 also possesses a nuclear localization sequence. VirD2 and virE2 then help the complex "home-in" on the nucleus and penetrate it. Once inside the nucleus, the T-strand integrates within the plant DNA and is converted into a double-stranded form. Details of the latter process have not yet been completely elucidated. A simplified schematic representation of the fate of the T-DNA, from *Agrobacterium* to a plant cell, is given in figure 3.5. Full details of the *Agrobacterium* system can be found in the recent review by Hansen and Chilton (1999).

Cell Wall Synthesis

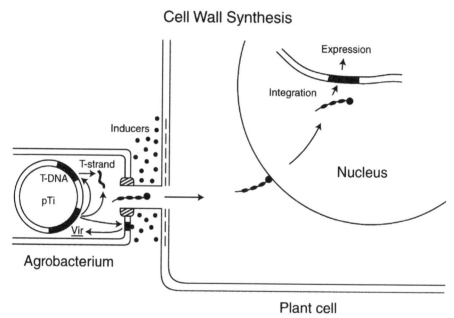

FIGURE 3.5.

Transfer of the T-DNA from *Agrobacterium tumefaciens* to plant cells. Inducers (e.g., methoxyphenols) are represented by *small dots* diffusing throughout the space separating the two cells. An inducer/Vir A complex is shown at the level of the *Agrobacterium* membrane. *Arrows* inside *Agrobacterium* depict the Vir cascade, including the synthesis of enzymes necessary to produce the T-strand, as well as Vir E2 coating the T-strand and VirD2 bound to its 5'-end. Exactly how the *Agrobacterium* pilus connects with the plant cell cytoplasm is not known. There is now evidence that Vir E2 can also travel on its own from *Agrobacterium* to the plant cell (see Hansen and Chilton 1999).

Thus, we have now reached the end of the crown gall story. This story will remain one of the great achievements of molecular genetics and has broken the myth that gene transfer between kingdoms is impossible. This type of transfer is undoubtedly rare today, but it may have been more frequent early in evolution. The unusual interactions between *Agrobacterium* and plant cells then pose the question of the origin of the T-DNA. The features of its genes (promoters, polyadenylation of its mRNAs) are definitely eukaryotic, although T-DNA lacks the introns that characterize many eukaryotic genes. Could it be that these are eukaryotic genes captured by *Agrobacterium* in the distant past? Quite possibly. At any rate, T-DNA genes are extremely weakly transcribed in *Agrobacterium*, as though they were not really "meant" to be functional in that environment.

As the solution of the crown gall problem was approaching, other events were taking place in two of the main groups responsible for these advances. One was the progressive dissolution of the Schell/Van Montagu combined team, as Schell had been offered a position at the Cologne Max-Planck-Institute. Jef Schell could not possibly be in Ghent and Cologne at the same time forever; the split was unavoidable. The Cologne group continued to do work with plant genetic engineering, solving both fundamental and practical problems. For a while, the Ghent team, under the direction of Marc Van Montagu alone, continued basic crown gall research and eventually ended up forming a company, Plant Genetic Systems (PGS), also devoted to practical genetic engineering. PGS is now part of AgrEvo, the agrochemical giant.

In 1979, Mary-Dell Chilton had been hired by Washington University in St. Louis, where she became a professor, and in 1983 she was recruited by another industrial giant, Ciba-Geigy (now Novartis) in North Carolina, to found a plant biotechnology research laboratory. On the other hand, Gene Nester has continued to elucidate the identity and role of the *vir* genes and still does *Agrobacterium* genetics in Seattle. He was chairman of his department until recently. Milt Gordon, also still at the University of Washington, turned his attention to the successful genetic transformation of poplar trees for use in xenobiotics detoxification, including the degradation of chlorinated aromatic pollutants (Perkins et al. 1990).

Victory: First Transformation of Plants

Agrobacterium is thus able to transfer some of its DNA to plants cells (the pTi T-DNA) with the help of the Ti plasmid *vir* genes. However, the transferred T-DNA harbors *onc* genes that are of no interest whatsoever to either the recipient plant or humans, at least in agronomic terms. Nevertheless, the fact that transposons, used for mapping T-DNA genes, were found integrated in the crown gall genome, "piggy-backing" on the T-DNA, immediately suggested that the T-DNA was a potential vector to introduce any gene into plant cells (Van Vliet et al. 1980). Surely, such engineered crown gall tissues were not much more interesting than regular crown gall (their only advantage was that they contained a bacterial transposon, hardly an advantage), but this showed that the *principle* of using the T-DNA to ferry foreign genes into plant cells was viable.

The trick, then, was to design a way to remove the oncogenes from the T-DNA and replace them with useful genes, for example genes of agronomic importance. Another challenge was to identify plants that had acquired a foreign gene. Of course, with fully virulent A. *tumefaciens* this had been easy: One had only to look for the appearance of a crown gall tumor. But in the total absence of oncogenes, crown gall would not form; how then would one tell the difference between a normal plant that had acquired a foreign gene and one that had not? In other words, how could one know that the gene transfer experiment was successful, since a transgenic plant would look exactly like a nontransgenic one? This is where the idea of genetic selection intervened again: Selectable marker genes were needed to enrich for and prove the transformed nature of a plant. In the next sections, we will see how the T-DNA from *Agrobacterium* was "disarmed," and what selectable genes (markers) were used to identify plant cells and eventually plants with introduced genes.

Let us first consider selectable markers. Antibiotic resistance had long been used in bacterial genetics to select for transformation events. The easiest way to know whether a particular technique would work to introduce DNA into a certain type of bacteria was to perform transformation experiments with plasmids carrying a gene that determined resistance to a given antibiotic. If the cells picked up the DNA, they would become resistant to this antibiotic and proliferate in its presence. Not so for the cells that had not incorporated any of the plasmid DNA: They would be killed by the antibiotic in the medium. This technique thus allowed differentiation between the transformed and the nontransformed cells. But what about plant cells? It turns out that they, too, are sensitive to a number of antibiotics, two examples of which are kanamycin (or G418) and chloramphenicol (Ursic et al. 1981; Lurquin and Kleinhofs 1982). Both chemicals kill plant cells by interfering with protein synthesis and, what's more, there exist bacterial genes (*neo* for kanamycin and *cat* for chloramphenicol) that can convert them into innocuous (for prokaryotes) derivatives by phosphorylation or acetylation. If one could transform plant cells with these genes, assuming the metabolites of kanamycin and chloramphenicol had become nontoxic, the cells would survive in the presence of antibiotics, and one would know that transformation worked. Also, the enzyme coded for by these resistance genes could be assayed fairly easily, allowing quantitative measurement of foreign gene expression. The *neo* gene is now universally used, whereas the *cat* gene no longer is because of its rather poor selectivity and the existence of many "escapes." Later, screenable markers such as the *gus* (β-glucuronidase) gene from *E. coli* and the *gfp* (green fluorescent protein) from jellyfish made it easy to visualize transformed plant cells, because they imparted a blue (*gus*, after incubation with the chromogenic substrate X-glu) or fluorescent green (*gfp*) color to transgenic cells harboring them (Jefferson et al. 1987; Leffel et al. 1997).

Of course, the selectable and screenable marker genes from bacteria and jellyfish (or any other nonplant source) could not be used directly in transformation experiments, because their transcriptional signals would not be recognized in plants. These markers had to be cloned and engineered so as to be under the control of plant promoters to be expressible. Genes consisting of promoters (and transcription terminators) controlling a foreign coding sequence, such as a bacterial one, are called chimeric, after the mythical composite beast. Plant-expressible chimeric *cat* and *neo* genes were constructed by cloning their coding sequences inside the T-DNA (with its *onc* genes), under the control of the *ocs* (octopine synthase) or the *nos* (nopaline synthase) promoter. Results with these chimeric selectable genes were published in 1983 by three independent groups within a period of four months (Herrera-Estrella et al. 1983; Bevan et al. 1983; Fraley et al. 1983). The Schell/Van Montagu team beat the others to print, but this year also saw the first show of force of the Monsanto plant genetic engineering group, a very serious challenger that, within no time at all, would leave a deep mark in the field. The reader will certainly notice the 14 authors on their first paper, an ominous sign. The third group to announce positive results with a selectable marker was the Mary-Dell Chilton laboratory from Washington University in St. Louis. All three groups had simultaneously disclosed evidence for success at the Miami Winter Symposium in January 1983.

At this point, the chimeric marker genes were introduced into plant cells via fully oncogenic Ti plasmids and, of course, crown gall was produced. However, and this was the point of these experiments, the marker genes worked. One could select for transformed plant cells without using the oncogenic trait. For the first time, it was convincingly demonstrated and independently confirmed that foreign chimeric genes could be introduced and function in recipient plant cells. Thus, 1983 was a turning point for plant transformation studies. Within weeks after their landmark *Nature* article (Herrera-Estrella et al. 1983) was published, the Schell/Van Montagu group reported the production of green plants, no longer tumors, harboring foreign genes (Zambryski et al. 1983). For this, it was necessary to eliminate the *onc* genes from the T-DNA. Mary-Dell Chilton's team had earlier used an indirect approach to generate transgenic plants, one that did not involve the physical removal of *onc* genes from the Ti plasmid. Instead, their transgene (a yeast alcohol dehydrogenase gene) had been inserted within the cytokinin-producing locus of a nopaline type of pTi (for details, see the next section).

Actual removal of oncogenes from the T-DNA was a feat of beautiful applications of recombinant DNA technology, whose details are much too intricate to be described here. Basically, the left and right portions of the T-DNA, containing the intact *nos* gene on the right side as well as the 25-bp borders on either side, but devoid of *onc* genes, were cloned in an *E. coli* plasmid, pBR322. After transformation into *A. tumefaciens*, this recombinant plasmid underwent a double-crossover with the resident, virulent Ti plasmid, with the result that the internal portion of the T-DNA (containing all the *onc* genes) was deleted and replaced with the *E. coli* vector, pBR322. This *Agrobacterium* strain thus now contained a pTi with fully functional *vir* genes, and a T-DNA still equipped with its borders, but basically consisting of pBR322 and the *nos* gene. This disarmed Ti plasmid was named pGV3850 (where GV stands for Ghent vector). *Agrobacterium* carrying this engineered pTi did not, of course, produce crown gall in infected plants. However, tiny masses of cells growing at the inoculation point were noticed, harvested, and regenerated into mature green plants by the addition of phytohormones. These plants were shown to produce nopaline (proof that the *nos* gene was present), and also Southern blots demonstrated the presence of disarmed T-DNA, equipped with its border fragments *and* complete pBR322 sequences. These plants now contained genes of bacterial origin, as claimed in chapters 1 and 2, but this time it was for real.

The Monsanto team confirmed these results a few months later, using a similar strategy involving a disarmed pTi, although it was constructed differently. This made sense: Monsanto was not in it simply for doing basic research. Being a company, they had to design their own transformation vectors (called pMON) for patenting purposes, something they did well. In addition, their report included evidence that a chimeric gene, a *nos-neo* construct, was integrated into the target plant genome via disabled T-DNA transfer and expressed (Horsch et al. 1984). Their plants had become resistant to kanamycin. Moreover, the foreign gene was transmitted to the progeny of the transformed plants by selfing. The foreign gene was thus inheritable. The Ghent/Cologne group had in fact only demonstrated foreign gene (pBR322) transfer and integration, but they could not show biological activity because the pBR322 genes were not chimeric and hence not plant-expressible. Also, they had not demonstrated that the foreign sequences could be transmitted by sexual crossing. They did confirm Monsanto's results almost immediately, however. Another article by the Monsanto group published about a year later (Horsch et al. 1985) established what would later be known as the "leaf disc cocultivation technique," which became enormously popular. The principle was quite simple: It consisted of excising leaf discs from the target plant using a paper punch, and incubating these leaf pieces with a suspension of engineered *A. tumefaciens* containing chimeric genes in its disabled T-DNA. These leaf pieces were then cultured on medium favoring shoot regeneration, in the presence of a selective antibiotic such as kanamycin (if *neo* was present in the T-DNA as a selectable marker). Rooted shoots growing in this medium would be transgenic because kanamycin would prevent growth of nontransformed plants by inhibiting root formation.

The concept of cocultivation of plant cells with *Agrobacterium* had actually been demonstrated many years earlier. As we saw before, *Agrobacterium's vir* genes are activated by the release in the outside world of compounds such as acetosyringone, building blocks of plant cell walls. This was unknown in 1978, when Laszlo Márton, from the Biological Research Center in Szeged, Hungary, went to Leiden, joining the Schilperoort group to do some work on plant transformation. Being a plant protoplast specialist, he thought that it might be possible to incubate cell wall–regenerating protoplasts with *Agrobacterium* and get transformation at the cellular level rather than at the whole-plant level, which involves thousands of cells. The idea here was to deal with single-cell transformation and study individual transformation events (as they are now called) rather than looking at a bulk phenomenon. He was right: Isolated protoplasts were transformable by *Agrobacterium*, meaning that a whole, differentiated plant was not necessary for *A. tumefaciens* infection (Márton et al. 1979). We now know that cell-wall synthesizing protoplasts do excrete significant amounts of acetosyringone, one of the requirements for susceptibility to T-DNA transfer and integration. It was later demonstrated that plant cells in established tissue culture, never protoplasted, can also be transformed by *Agrobacterium* (Müller et al. 1984; An 1985). It is now well known that many plant species can be transformed by *Agrobacterium* at the tissue culture level. Many types of actively dividing plant cells in tissue culture release acetosyringone in the growth medium. Genetic engineers often add this substance to the cocultivation medium "for good measure."

Transformation of plants via *Agrobacterium*-mediated gene transfer is now routine in many plant species. There are, however, two possible vector systems to achieve that goal. One of them is the pGV3850 system, and the other approach relies on the idea that the *vir* genes and the T-DNA (artificial or not—i.e., comprised of chimeric genes or not) need not be on the same plasmid (Hoekema et al. 1983; DeFramond et al. 1983).

In the pGV3850 system, most of the T-DNA consists of the *E. coli* cloning vector pBR322. This cloning vector and its derivatives are widely used to clone any gene in *E. coli*. Thus, once a gene of interest has been cloned, recombinant pBR322 can be introduced by transformation, as a suicide plasmid, into *Agrobacterium* harboring pGV3850. There, it will undergo crossing over with pTi, precisely within the T-DNA, inside the pBR322 region of homology. As a result, the foreign gene originally present in the pBR322 cloning vector will become integrated within the T-DNA and will be transferred to plant cells infected with this *Agrobacterium* strain (fig. 3.6). This indirect technique to introduce foreign genes into the T-DNA of pGV3850 is mandated by the large size of the disabled Ti plasmid; it is much too large to be manipulated *in vitro*, as its large size precludes the existence of convenient, single restriction endonuclease sites within its T-DNA. One major disadvantage of the pGV3850 system is that the inserted foreign gene is put into the plant genome flanked by two copies of pBR322, which constitute a large amount of extraneous, useless DNA.

Direct cloning of foreign genes within a disabled T-DNA became possible with the establishment of binary vector systems, in which the *vir* genes are located on a helper plasmid (such as pAL4404), itself devoid of all T-DNA sequences. The other member of the pair is a small plasmid containing the left and right borders of the T-DNA, no *onc* genes, and an antibiotic resistance marker for selection of plant transformation events. The left and right borders are separated by a multiple cloning site sequence containing several unique restriction endonuclease sites. The small size of this plasmid permits the direct cloning of foreign sequences between the left and right borders. When both the helper plasmid and the cloning vector are put together in the same *Agrobacterium* cell, the trans-acting *vir* functions will process the engineered T-DNA and transfer it to plant cells (fig. 3.7). The first practical vector was developed by Michael Bevan in 1984 (Bevan 1984), and several additional useful binary vectors of the pGA series were created by Gene An et al. (1985, 1986) a little later. Today, the pGV3850 co-

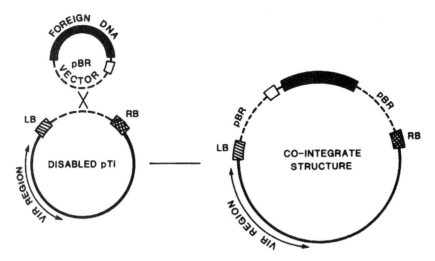

FIGURE 3.6.

Cointegrative vectors for the transformation of plants. An *Agrobacterium* strain containing a disabled Ti plasmid (with all oncogenes deleted) is transformed with a pBR (or pUC) type of plasmid harboring the foreign gene (conjugation can also be used to transfer the recombinant vector into *Agrobacterium*). The open box represents a marker selectable in both *E. coli* and *A. tumefaciens*. Because pBR and pUC vectors cannot replicate in *Agrobacterium*, transformants or exconjugants carrying the selectable marker are a result of homologous recombination between the vector itself and the vector portion present between T-DNA borders in pTi. The foreign gene (plus two copies of vector sequences) is thus incorporated between the T-DNA left and right borders (LB and RB) and can be transferred to plant cells. (From Lurquin, P. F. 1987. Foreign gene expression in plant cells. *Prog. Nucl. Acid Res. Mol. Biol.* 34:143–188, figure 2. Copyright Academic Press. Reprinted by permission of the publisher.)

integrate approach has become obsolete; all second-generation plant transformation vectors belong to the binary category, thanks to their greater ease of manipulation.

The foreign genes used during the birth of plant transgenesis from 1982 to 1984 were invariably antibiotic resistance markers. Why was this, and why did researchers not use other genes, such as those coding for disease resistance or other useful traits? The reason is simple: Such useful genes had not been cloned and were not available. But more important, in those days, the *feasibility* and *reproducibility* of plant transformation had to be ascertained. For this, genes imparting unequivocal phenotypes to the transformed plants had to be used, and antibiotic resistance genes were a perfect

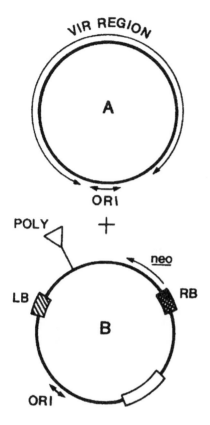

FIGURE 3.7.
Binary vectors for the transformation of plants. The *A. tumefaciens* resident A plasmid is a severely deleted pTi, devoid of oncogenes but containing an intact *vir* region and its origin of replication. The B plasmid is the foreign gene-cloning vector equipped with a wide host range origin of replication (for replication in both *E. coli*, where cloning is performed, and *A. tumefaciens*). The B plasmid also harbors an antibiotic resistance gene for selection of B-containing *A. tumefaciens* cells after transformation (*open box*). The foreign gene can be cloned into a polylinker region located between the left and right borders (LB and RB) of the disarmed T-DNA. Such binary vectors also contain plant-selectable genes (such as *neo*) and/or plant-screenable genes such as *gus* or *gfp*. (From Lurquin, P. F. 1987. Foreign gene expression in plant cells. *Prog. Nucl. Acid Res. Mol. Biol.* 34:143–188, figure 3.)

choice; in the presence of lethal doses of an antibiotic, only transformed plant cells and plantlets would survive. Clearly, many genes of agronomic importance have been used to transform plants since these pioneering days (see chapter 5).

Inheritance of Transgenes by Progeny of Transformed Plants

We saw in chapter 2 that classical genetics results are crucial in determining the reality of biological phenomena involving DNA. It really is not enough to demonstrate presence and expression of a foreign gene in an organism; the study of the transmission of this foreign gene to the offspring is equally important, for sometimes unhappy surprises may occur. A brief allusion to transgene heritability was made in the preceding section. This section describes in greater detail some of the crucial experiments that first demonstrated foreign gene transmissibility in plants. Genetic experiments with plants regenerated from tumor tissues of tobacco were performed by the Ghent/Cologne group (associated with researchers from the University of Brussels) as early as 1981 (Otten et al. 1981). As described previously, crown gall cells are extremely refractory to regeneration into green plants. Further, in 1981, the nature of the *onc* genes was still poorly understood and, of course, disabled Ti plasmids did not exist. How then could anyone conduct genetic experiments, which require fertile plants, under those circumstances? The answer is, a good dose of luck and recognizing opportunity when it knocks. The Schell/Van Montagu team had produced a collection of *A. tumefaciens* mutants containing transposons in the T-DNA. The objective was to "knock out" the T-DNA genes and see what kinds of tumor phenotypes were observed after plant infection. It turned out that one of these mutants, pGV2100, caused the formation of green shooty tumors rather than an undifferentiated mass. Hundreds of these shoots were analyzed for the presence of octopine and found to be negative, but a single shoot was found to be octopine-positive and able to root, just like a normal plant. This plant was named rGV-1 by its creators and *wunderpflanze* (German for wonder plant) by those who had tried the same experiment without success. (The reference to Jef Schell's recent hiring by the Max-Planck-Institute was clear.)

Why was the *wunderpflanze* so special? First, of course, was its regeneration from a crown gall transformation event. Next was the fact that it produced octopine, meaning that at least the *ocs* gene had been transferred by the T-DNA. Southern blot analysis demonstrated that, in fact, the *ocs* gene was just about all that was left in this plant of the whole T-DNA. No *onc* genes were found, explaining why regeneration was possible from an octopine-positive shoot. What the researchers had stumbled on was a very infrequent event in which only a small part of the T-DNA had been integrated in the transformed cell that gave rise to rGV-1. Basically, a big deletion had been responsible for the elimination of the *onc* genes. Deletions are now known to be fairly commonly observed in T-DNA-mediated transformation, but this was new at the time and it was a really big deletion. Having shown that this plant contained and expressed the foreign *ocs* gene, the team started classical genetics experiments without, however, providing any statistical information. The calculations that follow are mine. When selfed, rGV-1 produced a 3:1 ratio between octopine-positive and octopine-negative plants, with a χ^2 value of 2.97 at $p = .05$. Fine. In crosses with the wild-type, 1:1 ratios were observed in two experiments, with χ^2 values of

3.34 and 0 at p = .05. Borderline in one case, but excellent in the other. Finally, anther cultures (it is possible in some cases to regenerate whole plants from pollen grains) produced octopine-positive and octopine-negative haploid plants in a 1:1 ratio with a χ^2 of 3.17 at p = .05. Fine too. All three experiments concurred: The T-DNA-mediated transfer of the *ocs* gene produced plants hemizygous for the gene (heterozygous, if one prefers), and transmission to the offspring was Mendelian. Further, the *ocs* gene was dominant, which was expected because this gene was *added* and thus had no counterpart in the recipient plant. This was excellent news for those contemplating stable, inheritable gene transfer into plants.

These results were later extended and confirmed by others (Barton et al. 1983; Memelink et al. 1983). Work done by the Chilton group in St. Louis, in collaboration with Andrew Binns at the University of Pennsylvania (Barton et al. 1983), is particularly instructive regarding the process of scientific research and the usefulness of good collaborations. This group had been the first to produce transgenic plants without having to remove the oncogenes from the T-DNA. This is how it worked. Ken Barton, a postdoctoral fellow in Chilton's lab, had engineered a yeast alcohol dehydrogenase (ADH) gene into the T-DNA of a nopaline *Agrobacterium* strain. He found that this strain would not make tumors able to grow on hormone-free medium. At this point, this strain could have been discarded or stored and forgotten. Rather, it was sent to Binns, a tobacco tissue-culture expert, who found out what the problem was. It turned out that the transformed tobacco cells could grow in the absence of auxin but required cytokinin. Thus, the inserted yeast gene had inactivated the T-DNA cytokinin locus. At this point again, these tumors could have been thrown out because the problem of absence of growth on hormone-free medium had been solved. On the contrary, these cultures were kept in the growth chamber and, after a few weeks, started making shoots. Tests for the presence of nopaline turned out positive and Southern blot analysis showed the presence of the original T-DNA with the ADH gene. Here again, these shoots could have been discarded, because shoot regeneration from nopaline-producing teratomas was not new. The shoots were not disposed of and a few weeks later, they started to produce roots. This was news, because it was known that ordinary teratomas simply did not root. This new phenomenon could have been attributed to a deletion of the T-DNA, as in the example of the wonder plant. This proved not to be the case; the plantlets produced nopaline abundantly and still contained the original T-DNA. This meant, in effect, that insertion of the yeast gene had disarmed the T-DNA, and that a completely foreign gene was now present in a regenerated tobacco plant (the ADH gene was not expressed, however)! It is likely that once the cytokinin locus had been inactivated by the ADH gene insertion, the low-level expression of the auxin genes (which is the norm with nopaline T-DNAs) did not interfere with plant regeneration. The same results were obtained with an *Agrobacterium* strain containing this time a chimeric *neo* marker inserted at the same location in the T-DNA. *Neo*-containing plants flowered and set seed, and their F1 progeny was found to contain an intact, engineered T-DNA. The foreign gene was thus inheritable and there was even some evidence of segregation. To celebrate the occasion of this discovery, one of these regenerated transgenic plants appeared on the cover of the issue of *Cell* in which the report was published. The moral is that scientists should be in their lab when opportunity knocks on their door!

It was later found that complicated T-DNA integration patterns could lead to non-Mendelian segregation ratios. The notions of cosuppression and gene silencing derive from these observations (chapter 5). In an extensive study with *Petunia*, the Monsanto group identified engineered T-DNA integration events on three of the seven *Petunia* chromosomes. Also, they showed that multiple T-DNA integration events were possible, including integration in one spot of T-DNA units repeated in tandem (Wallroth et al. 1986). This is indeed pretty much what we know today: T-DNA integrates within recipient plant DNA in a largely random fashion, can integrate as a dimer or sometimes higher-order multimer, and multiple copies of the T-DNA can also be found in different chromosomal locations in the same transformed plants.

A good quantitative idea of the frequency of simple integration patterns (which lead to simple Mendelian inheritance) compared to more complicated ones was provided by the large-scale studies of Deroles and Gardner (1988a, 1988b), which demonstrated that, as usual, things were much more complicated than originally thought. Their statistical calculations were impeccable. They transformed petunias with a disabled T-DNA-borne *neo* gene and isolated transformants growing in the presence of kanamycin. Inheritance patterns were studied by selfing and testcrossing (see appendix 3), and outcomes of the crosses could be separated into seven categories. The most numerous one (36/104) had plants segregating 3:1 on selfing, and 1:1 on testcrossing. These ratios are consistent with the presence of a single *neo* allele: These plants were hemizygous for the transgene. The next class (6/104) segregated 15:1 and 3:1, ratios typical of two unlinked genes, and thus these plants contained two independently integrated copies of the *neo* gene. The third class (1/104) generated 100 percent kanamycin-resistant progeny in both crosses and was thus homozygous for the *neo* gene. Finally, a fourth class (6/104) generated a 2:1 ratio by selfing, *but* it generated a 1:1 ratio in testcrosses, strongly suggesting that the T-DNA had inserted into an essential region of the genome, and that a homozygous condition was lethal (hence the loss of 25 percent of the progeny after selfing and a 2:1 ratio). These four classes all conformed to Mendelian patterns of inheritance.

However, over 50 percent of the transgenic plants (55/104) did not demonstrate Mendelian inheritance. Twelve out of 104 of these plants had no kanamycin-resistant progeny at all (class 5), but only two of those had lost the *neo* gene. Class 6 transformants (33/104) showed ratios less than 3:1 and less than 1:1, whereas class 7 (10/104) showed variable ratios in independent crosses. The existence of plants in these last three classes cannot be explained in Mendelian terms. The authors invoked epigenetic factors (such as DNA methylation) to explain these anomalies. Also, plants in the non-Mendelian categories harbored multiple copies (three to six) of the T-DNA, a situation potentially leading to cosuppression (gene silencing resulting from allele multiplicity). No cases of cytoplasmic inheritance were found.

Multiple copy integration, gene silencing, and non-Mendelian inheritance are not restricted to "laboratory plants" such as tobacco, petunia, and *Arabidopsis*. We have found similar effects in *Agrobacterium*-transformed wild potatoes (Cardi et al. 1992). Thus, plant genetic engineering is not simply a matter of "taking that gene and throwing it in there," as lay audiences too often believe. There is indeed a long, treacherous, and expensive path between the primary transformant appearing in a Petri dish and a field full of transgenic crop plants.

Conclusions

There is no doubt that the elucidation of crown gall disease will remain one of the great achievements of plant molecular biology. This is of course not because crown gall is of major agronomical importance (it is not), and it is in fact still incurable. Rather, not only has the study of this disease revealed fascinating aspects of gene transfer between plants and a bacterium, but it has also made possible the genetic manipulation of higher plants through recombinant DNA methodology for basic and applied purposes. The history of this endeavor shows again that fundamental research can have completely unpredicted consequences. Without an understanding of the molecular nature of crown gall and its application to plant transformation, it would not have been possible to unravel many of the genetic pathways known today. Similarly, the rapid pace at which plant biotechnology is now advancing would have been impossible without these basic discoveries. A short anecdote can be used to illustrate the fundamental nature of crown gall research. Jef Schell once commented at a scientific meeting, "Nester and Gordon are the only two people I know who could convince university authorities to do plant research and build a greenhouse in a hospital!" And indeed, the Nester and Gordon labs are located at one end of the enormous University of Washington Medical School building, hardly a place to conduct applied agronomical research, but definitely a good one for basic molecular biology.[2]

CHAPTER 4

Direct Gene Transfer

By 1979, there was still no solid, published evidence that DNA-mediated transformation of plants was possible. This approach to plant genetic engineering is of course different from the *Agrobacterium*-mediated transformation system, in which a live vector (the bacterial cell) is necessary to achieve DNA transfer. It should also be remembered that in 1979, there was as yet no evidence that the *Agrobacterium* T-DNA would turn out to be an excellent vehicle for plant transformation. Therefore, DNA-mediated transformation (or direct gene transfer, as it would become known later) was an approach to an unsolved problem—a competing avenue to create engineered plants. What was happening with other eukaryotic systems at that time? Enormous progress had been made with yeast and animal cells, and there was no longer any doubt that DNA-mediated transformation was a reality in these systems. For example, transformation of yeast had been established in 1978 (Hinnen et al. 1978) and the construction of yeast–*E. coli* shuttle plasmids based on the 2-micron nuclear plasmid had been published that same year (Beggs 1978). By 1979, mouse cells had been transformed with eukaryotic and prokaryotic DNA, and there was no doubt that totally heterologous sequences such as pBR322 (a bacterial cloning vector) and ϕX174 (a phage) could be integrated and replicated along with mouse chromosomal DNA (Wigler et al. 1979). Transformation of mouse embryos by microinjection with recombinant plasmids was published in 1980 (Gordon et al. 1980). In light of this, plant transformation was a poor prospect indeed—so poor in fact that a general review of the field of genetic transformation that came out in 1981 did not even make a single allusion to plant systems, and wisely so (Smith and Danner 1981). Nevertheless, many review articles were published by scientists interested in the DNA transformation of plants and, of course, these were totally speculative. An author of one of these reviews commented verbally, "There are more review articles out there than genuine data." However, the waiting was almost over.

Development of DNA Uptake Systems

It had become clear by the mid 1970s that incubating seeds with purified DNA was not the way of the future if one wanted to transform plants by direct gene transfer. Even though good selectable markers for the singling out of transformation events did not exist, several researchers were confident that they would be available one day, as recombinant DNA technology was making great advances. Such recombinant DNA molecules could then be used later in transformation experiments once uptake conditions had been established. Further, it was thought that isolated plant cells, rather than whole plants, would be much easier to manipulate under controlled conditions. However, it was also believed that the thick cellulosic cell wall of plant cells would completely prevent entry of DNA. We now know this not to be entirely the case, although the cell wall does present a significant barrier to DNA uptake. Therefore, many labs embarked on the study of DNA uptake by plant protoplasts (fig. 4.1), cells rid of the cell wall by enzymatic digestion.

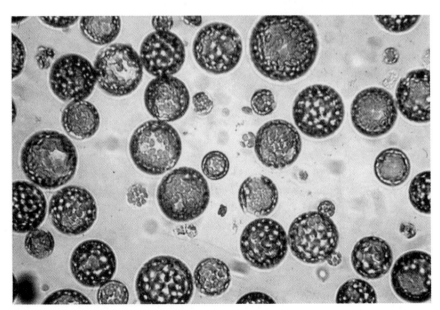

FIGURE 4.1.
Tobacco mesophyll protoplasts.

There was a good precedent for this approach. Protoplasts had been shown to be permeable to viral RNA molecules and able to replicate and translate them for the synthesis of complete viral particles. If RNA could penetrate protoplasts, so could DNA, in all likelihood. Also, protoplasts were "hot" in those days, as they had allowed plant cell genetics to join the ranks of bacterial and cultured animal cell genetics in terms of ease of manipulation *in vitro*. Moreover, plant protoplasts from some species were totipotent and could be coaxed to regenerate into whole plants from single cells, meaning that rare genetic events could be identified and studied in a whole, fertile, regenerated organism.[1] This has not yet been achieved in the animal kingdom.

Aoki and Takebe (1969) had published a convincing set of results demonstrating that pure tobacco mosaic virus (TMV) single-stranded RNA could penetrate tobacco protoplasts, replicate, and become encapsulated in the virus coat protein. Several years later, the same authors published a detailed study of the kinetic aspects of TMV RNA replication in protoplasts (Aoki and Takebe 1975), leaving no doubt that naked RNA could be biologically expressed after uptake. One major conclusion could be drawn from these observations: A large nucleic acid could cross the membrane of protoplasts, but only if its negative electrical charge was at least partially neutralized. Indeed, DNA and RNA are both polyanions at physiological pH because of their negatively charged phosphodiester backbone, but so is the cell membrane. Thus, cell membranes and nucleic acids repel each other and uptake would not be expected. To circumvent this problem, these authors complexed TMV RNA with polycations (positively charged molecules) such as poly-L-lysine and poly-L-ornithine to neutralize the negative charges of the RNA. The trick worked, and the RNA was picked up by the protoplasts and expressed. Of course, viral RNA is translated and replicated in the cytoplasm of the cell and thus does not need to penetrate the nucleus to be expressed. Yet these experiments demonstrated that the cell membrane was permeable to macromolecules, and this alone was exciting.

The first group to understand the implications of these observations for the genetic engineering of plants was the Saskatoon, Canada, protoplast lab. Unfortunately, in 1972 plant protoplasts were still curiosities, attracting little notice, and their results were at a preliminary stage, basically showing DNA adsorption to membranes in the presence of polycations, but with no proof of intracellular absorption (Ohyama et al. 1972). Little did anybody know at the time that this Canadian team was on the right track.

What in fact are the prerequisites for DNA-mediated transformation of cells? The first hurdle of course is to ensure that DNA molecules will be taken up. This is not as trivial a phenomenon as it may seem. As mentioned, it is expected that the negative charge of DNA will be repelled by the negative charge of the cell membrane. In the case of mammalian cells, a clever trick was designed in which viral DNA was presented to recipient cells in the form of a very fine calcium phosphate–DNA coprecipitate, which was taken up by endocytosis (Graham and van der Eb 1973). This technique worked quite well and was used for many years to achieve the genetic transformation of a variety of mammalian cells. It is interesting to note that, somehow, DNA taken up by endocytosis made its way into the nucleus where it could be expressed. This method never worked well with plant protoplasts, which do not naturally perform endocytosis. On the other hand, the use of polycations, which allowed uptake and expression of viral RNA in the cytoplasm of plant protoplasts, was not necessarily a good indication that DNA could further end up in the nucleus. In fact, in spite of a positive report (Yamaoka et al. 1982), the uptake and expression of the circular, double-stranded DNA genome of cauliflower mosaic virus (CaMV) (which needs a nuclear step for replication and transcription) in protoplasts was not considered conclusive by most investigators. Therefore, contrary to what happened with mammalian cells, a model viral DNA uptake/expression system was never developed with plant protoplasts. In other words, no reliable biological expression system was identified to optimize uptake and expression of foreign DNA in plant protoplasts.

Under those circumstances, only two courses of action were possible: (1) Do nothing and wait for others to solve this problem, or (2) establish experimental conditions under which donor DNA is internalized, as evidenced by its physical presence (not its biological activity) inside protoplasts. Several research teams chose the latter option, and the years from 1976 to 1983 saw a flurry of activity in this area. The first serious efforts were published by Suzuki and Takebe (1976) from Tokyo and Chiba, Japan, using radiolabeled coliphage fd single-stranded DNA as donor, and tobacco protoplasts as recipient. This DNA was used presumably because it is single-stranded and has roughly the same molecular mass as TMV RNA and thus mimics it. Their data showed that about one third of input DNA became irreversibly bound (DNase-resistant) to protoplasts, and 70 percent was recovered in a cytoplasmic fraction and the rest in a fraction containing nuclei, chloroplasts, and mitochondria. The degree of polymerization of protoplast-associated donor DNA was somewhat decreased, but 30 percent of these molecules cosedimented with intact fd DNA in a sucrose gradient. As expected, polycations (in particular poly-L-ornithine combined with zinc ions) had a dramatic positive effect on DNA uptake. Interestingly, these conditions of incubation led to high infectivity values when TMV RNA was used, implying that this RNA must have been taken up efficiently and, likewise, fd DNA presumably was.

It was in 1976 also that the first versatile plasmid-cloning vehicle (pBR313) was developed by Francisco Bolivar and Ray Rodriguez in the lab of Herbert Boyer at the University of California, San Francisco. It was then easy to become convinced that sooner or later, plasmid vectors would be used to transform plants with cloned genes, possibly even with antibiotic resistance markers for selection purposes (Lurquin 1976). Basically using the conditions developed by Suzuki and Takebe (1976), it was shown by others that about 50 percent of absorbed donor pBR313 DNA became associated with cowpea protoplast nuclei after a short incubation time. Internalized, polymerized plasmid molecules were on average cut by endogenous nucleases only twice during the process, and, unsurprisingly, no integration of the foreign DNA was detected by hybridization analysis (Lurquin and Kado 1977). These results were quickly confirmed and extended to other protoplast species by at least five independent groups, some using plasmid DNA (Hughes et al. 1977; Hughes et al. 1979; Owens 1979), others using *E. coli* chromosomal DNA (Liebke and Hess 1977; Uchimiya and Murashige 1977) or phage λ DNA (Suzuki and Takebe 1978). Interestingly, Owens (1979) found no evidence for kanamycin resistance in tobacco calli incubated with a ColE1-kan plasmid. This should not surprise us today since this plasmid's *neo* gene was under the control of prokaryotic transcriptional signals.

From the dozen or so articles published on the topic of DNA uptake into plant protoplasts, three findings emerged. First, it was necessary to protect the donor DNA in the incubation medium, lest it be rapidly degraded by nucleases released by broken protoplasts (and sometimes present in the crude enzyme preparations used to isolate the protoplasts from plant tissues). Protection could be achieved in a number of ways, either by complexing the DNA with polycations or polyethylene glycol (PEG), or by increasing the pH of the incubation medium (Lurquin and Márton 1980). Second, significant uptake was observed only when the protoplast membrane was "permeabilized," which, interestingly, could be done using the previously mentioned agents, as demonstrated in studies involving viruses or viral RNA. Finally, several independent laboratories agreed that polymerized donor DNA could reach the nuclear compartment. This, of course, still did not mean that any of this DNA could be expressed, given the fact that no plant expressible genes had yet been cloned at the time. Nevertheless, DNA uptake studies in protoplasts continued, with the objective to maximize internalization of donor molecules while keeping them as intact as possible.

It had been known since the mid 1970s that polycations, PEG, and high pH (above 10.0), in the presence of calcium ions for the latter two, induced plant protoplast aggregation and fusion. Also, as we have seen, these conditions enhance uptake of nucleic acids by protoplasts. It was therefore natural to start thinking about ways to somehow encapsulate donor DNA in some kind of lipidic membrane and fuse this complex with recipient protoplasts. In this way, DNA would presumably be very well protected from nuclease attack in the incubation medium, and delivery into the cytoplasm of the protoplasts would be achieved by the action of a fusogen, such as PEG. Such lipidic vesicles did exist and they were called liposomes. Further, it had been demonstrated in 1978 that liposomes were able to trap polymerized DNA (Hoffman et al. 1978) and that encapsulated mRNA was biologically expressed in mammalian cells exposed to liposomes (Dimitriadis 1978).

In 1979, the first report showing plasmid DNA trapping in liposomes and fusion with plant protoplasts was published (Lurquin 1979). Experiments showed that radiolabeled plasmid DNA and DNA–fluorescent dye complexes could be efficiently encapsulated in large multilamellar liposomes (see appendix 5), which could be prompted to fuse at high frequency with protoplasts in the presence of PEG. Here, also, nuclear transfer of reasonably intact plasmid DNA was observed. Very similar results had been obtained independently, practically at the same time, by two research groups (Lurquin 1979; Dellaporta and Giles 1979). We will see that liposome-mediated nucleic acid transfer into protoplasts was to enjoy a few years of success but was eventually supplanted by simpler gene transfer methods.

First Hints

Again, showing DNA uptake by protoplasts is a far cry from demonstrating biological activity. Nevertheless, inspired by experiments done with radiolabeled plasmid DNA as described in the preceding section, two groups had obtained data deemed presentable at the Fifth International Protoplast Symposium held in July 1979 in Szeged, Hungary, on the effect of purified Ti plasmid DNA isolated from *A. tumefaciens*. It will be recalled that in 1979, the integration and expression of pTi T-DNA in crown gall cells had been well established, but there was still no evidence that pure, donor pTi DNA was sufficient to produce a crown gall phenotype if taken up by cells. Also, no pTi-based vectors existed at the time and there were no good clones of the whole T-DNA. Further, it was still unclear whether non-T-DNA genes were involved in, for example, its integration within plant genomes. This explains why the entire Ti plasmid was used as donor DNA in the experiments that will be described later. As chimeric, plant-expressible markers did not yet exist, the *Agrobacterium* Ti plasmid was the only possible choice to conduct biologically meaningful DNA uptake experiments.

Thus, the Leiden and Nottingham teams (both of them in collaboration with Belgian researchers but working independently) reported on the uptake and biological effects of pTi DNA incubated with protoplasts (Márton et al. 1979; Davey et al. 1979). In the first case, tobacco protoplasts were incubated with pTi DNA at pH 10.5 in the presence of calcium ions, whereas in the second, *Petunia* protoplasts were incubated with pTi DNA complexed with poly-L-ornithine. Treated protoplasts were then allowed to regenerate a cell wall and plated on hormone-free medium, which did not allow the division of normal cells but permitted division of cells transformed to a crown gall phenotype through pTi uptake and chromosomal integration. Hormone-independent clones were recovered in both cases, some of them displaying octopine dehydrogenase activity, a sole property of crown gall cells. Was that finally it? Was DNA-mediated plant transformation finally demonstrated? Well, not everybody was convinced at the time. This is because no Southern blots demonstrating the presence of T-DNA in these transformants were available in 1979 in time for the Protoplast Symposium. These blots followed just a few months later and confirmed that both the tobacco and *Petunia* clones were indeed genuine transformants. Thus, we can consider 1979 to be the (perhaps unofficial) birthdate of plant cells transformed by direct gene transfer. Sadly, these results were never published in a refereed journal as a full "package" by either group. Nevertheless, the first complete demonstration of DNA-mediated transformation of plant cells was published in 1982 by the Leiden group, using tobacco protoplasts, the Ti plasmid, and the fusogen PEG to promote uptake (Krens et al. 1982).

Liposome-mediated pTi transformation of protoplasts had already been observed in 1981 (Dellaporta 1981), however. Still, at that point, the problem was that transforming protoplasts with Ti plasmid DNA was not exactly plant genetic engineering yet. Granted, this work demonstrated once and for all that crown gall was akin to genetic, DNA-mediated transformation, and that *Agrobacterium* was now just a convenient (and still a much more efficient) way to deliver T-DNA to plant cells. It was by now quite obvious that the next step would be to try to transform plant protoplasts with chimeric, plant-selectable genes, once they were available.

The Finish Line

The incontestable credit for having published, in 1984, an ironclad demonstration of plant transformation by direct gene transfer, using non-pTi DNA, belongs to the Basel, Switzerland, group (Paszkowski et al. 1984). This team was composed of researchers versed in protoplast manipulation, classical plant genetics, and virus molecular genetics, a combination that proved to be highly synergistic. Their strategy was to clone a plant-selectable marker under the control of CaMV gene promoter and terminator sequences, and to incubate this transformation vector with totipotent tobacco protoplasts.

Cauliflower mosaic virus has a double-stranded DNA genome characterized by three single-stranded gaps possessing short single-stranded tails. The gaps are repaired and the tails trimmed after cellular uptake and uncoating, and the viral genome is then found in nuclei as a supercoiled minichromosome associated with histones. This structure is then transcribed by endogenous RNA polymerase II into a full-length, 35S RNA species, which is subsequently reverse transcribed by a virus-encoded reverse transcriptase (Pfeiffer and Hohn 1983). Virions are assembled in cytoplasmic inclusion bodies whose protein membrane is translated from a virus-encoded 19S mRNA species. Because purified CaMV DNA is infectious in whole plants, it had been hoped at one point that the whole CaMV genome could be used as a replicating plasmid-like entity that would sustain the transgene intracellularly. These hopes were shattered when it was demonstrated that inserting more than 250 bp of DNA abolished viral DNA replication (Gronenborn et al. 1981). This size limitation would preclude use of this vector for transformation with practically any gene (except for the unusually short *dhfr* gene, which codes for dihydrofolate reductase).

Nevertheless, from studies with CaMV DNA, two strong promoters, the 19S and 35S, were identified and cloned, together with their common terminator signal. The coding sequence of the *neo* gene was thus cloned between the 19S (also originally called gene VI) promoter and terminator and the construct inserted into the *E. coli* vector pUC8. The resulting plasmid was then introduced into tobacco protoplasts via the PEG technique, and selection of tranformants was achieved in the presence of kanamycin (Paszkowski et al. 1984). Five kanamycin-resistant tobacco clones were recovered, bringing the transformation frequency to 10^{-5}, based on the number of cells surviving the treatment. (fig. 4.2). Plants regenerated from these transformants continued to display the kanamycin-resistant phenotype and were shown to have the *neo* transgene integrated in nuclear DNA and to express it enzymatically. Genetic analysis showed Mendelian segregation of the trait in one plant but a more complicated pattern in another one. Interestingly, both regenerated plants harbored multiple copies of the transgene, albeit only a single intact copy was present. The earlier observations of Krens et al. (1982) had shown that direct gene transfer of pTi DNA resulted in the integration of DNA segments of variable length, and their subsequent experiments even showed integration of calf thymus DNA used as carrier (Krens et al. 1985). In fact, scrambling, duplication, integration of concatemers, and integration at multiple independent loci have been demonstrated in direct gene transfer experiments, regardless of the uptake technique used. Although *Agrobacterium*-mediated gene transfer can encounter similar complicating factors (see chapter 3), the situation is much, much worse in the case of direct gene transfer. But these were at the time mere details compared with the fact that, at last, DNA-mediated plant genetic engineering with completely foreign genes had been demonstrated.

In essence, the work of the Basel group put an end to the controversy that had surrounded the whole field of foreign DNA uptake and integration in plants since 1967. The road to success had been a long and tortuous one, but at last it was open. The achievements of the Basel team vindicated collectively all those who had not lost hope in this distant goal and had kept the field alive in spite of all the earlier negative press given it.

It is remarkable that the first evidence for chimeric, direct foreign DNA transfer was obtained only a year after the publication of *Agrobacterium*-mediated transformation. Even though, at that time, exactly how the T-DNA was transferred to plant cells was still unknown, it had become clear that once DNA was internalized in plant protoplasts, it had a good chance to become integrated and expressed, as demonstrated with naked pTi. Therefore, *Agrobacterium* itself was not necessary for transformation, and direct gene transfer was no longer an outlandish thought. Here, also, the demonstration of direct gene transfer in 1984 was a turning point announcing an era of applications rather than one of basic discovery. It should be noted, however, that the mechanism by which DNA becomes integrated after direct gene transfer is still unknown. Of course, this is also the case for T-DNA integration. Whether the two processes are similar or entirely different remains to be seen.

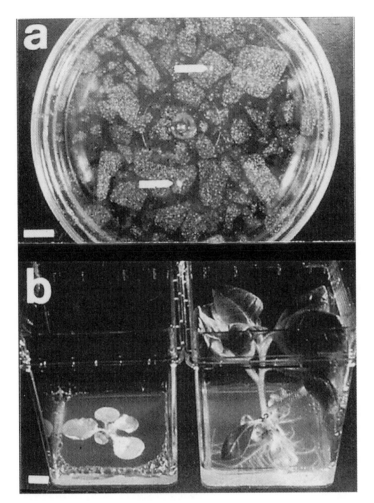

FIGURE 4.2.

Transformed tobacco cells clones and regenerated transgenic plants obtained for the first time by direct gene transfer of a chimeric selectable marker. (*a*) Protoplasts transformed with a *neo*-bearing plasmid incubated in agarose slabs in the presence of growth medium containing kanamycin. *Arrows* point at two transgenic colonies. (*b*) *Left*: Control nontransgenic tobacco plant dying in the presence of kanamycin (note the absence of roots). *Right*: Transgenic tobacco plant regenerated from a colony in (a) and growing in the presence of a lethal dose of kanamycin (note the presence of well-developed roots). (From Paszkowski, J., R. D. Shillito, M. Saul, et al. 1984. Direct gene transfer to plants. *EMBO J.* 3:2717–2722, figure 2. Copyright Oxford University Press. Reprinted by permission of the publisher.)

A Proliferation of Techniques

As described previously, protoplast transformation was first achieved with *Agrobacterium* pTi DNA, in the presence of either poly-L-ornithine, PEG, calcium ions at high pH, or liposomes. At first, and before the announcements of the Basel group in 1984, it seemed as if DNA entrapment in liposomes would be the way of the future, given the excellent protection they afforded against extracellular nucleases and their high frequency of fusion with protoplasts. Following initial results obtained with encapsulated plasmids (Lurquin 1979; Dellaporta and Giles 1979), several labs embarked on projects aimed at refining the delivery systems using radiolabeled DNA (Uchimiya and Harada 1981; Rollo et al. 1981; Matthews and Cress 1981; Lurquin and Sheehy 1982; Lurquin and Rollo 1983; Sheehy and Lurquin 1983) or viral RNAs (Fukunaga et al. 1981; Fraley et al. 1982; Rollo and Hull 1982; Christen and Lurquin 1983; Rouze et al. 1983) and, finally, demonstrating genetic transformation with a chimeric plant selectable marker (Deshayes et al. 1985; Caboche and Lurquin 1987). Quite certainly, not all groups agreed on the best lipid composition for the liposomes, or even the nature of the fusogen. But the strategy was the same in most cases: Large unilamellar liposomes worked best and PEG was quite adequate in promoting DNA or RNA transfer to protoplasts. Unfortunately, there was a significant disadvantage associated with this method: It worked quite well, especially with certain viral RNAs and viroids, but, with DNA, transformation frequencies were not any better than those obtained with the much simpler liposome-free PEG method used by Krens et al. 1982! Given the serious difficulties at the time to produce sterile liposomes and the numerous manipulations involved, this technique was eventually abandoned. Interestingly, "new generation" cationic liposomes that promote gene transfer at very high frequencies in animal cells have, to my knowledge, been used only once with plant protoplasts (Antonelli and Stadler 1990). In all likelihood, this is because protoplasts are no longer a necessary step in the production of transgenic plants via direct gene transfer. Moreover, regeneration of plants from protoplasts often results in somaclonal variation and is a more lengthy and expensive process than producing plants from intact cells. Thus, direct DNA transfer to protoplasts is not the most attractive approach today.

However, before the demise of protoplasts, one more technique based on their use was developed, practically in parallel, in the United States and Switzerland. It had been shown in 1982 that brief electric pulses administered to mouse suspension cells in the presence of DNA led to the expression of a transgene (Neumann et al. 1982). The technique became known as electroporation because an electric discharge causes the transient opening of pores in the cell membrane (appendix 6). These pores allow the import of macromolecules, including DNA. Gene transfer by electroporation was rapidly extended to many kinds of animal cells, plant protoplasts, fungi, and bacteria (Lurquin 1997) and is thus the most versatile transformation technique ever designed.

The first research team to apply this technology to the genetic transformation of plant protoplasts was Virginia Walbot's at Stanford University (Fromm et al. 1985). Their results were confirmed within weeks by the Basel group, which also showed stable integration and expression of a transgene whereas the Stanford team had shown only transient transgene expression, presumably without integration (Shillito et al. 1985). Soon thereafter, electroporation rapidly superseded all other techniques of gene transfer into plant protoplasts. However, a benign controversy then surfaced. Because the two groups used very different electrical parameters to achieve basically the same result, questions were raised as to which set of conditions worked better. Some groups sided with the low voltage and long discharge approach (Fromm et al. 1985), and others preferred to use high voltage combined with a much shorter time of discharge (Shillito et al. 1985). However, comparison between the results obtained by these two sets of conditions did not show a drastic difference at all. This was seen as a puzzle by some, who sometimes invoked different poration mechanisms brought about by the different conditions. There is in fact nothing of the sort. It turns out that, within the parameters consistent with cell survival and membrane breakdown, the short or long pulses at high or low voltage dissipate the same levels of electrical energy in the system and affect cell membranes equally (Lurquin 1997; Chen et al. 1999; Lurquin et al. in press).

The preponderance of electroporation lasted a few years and resulted in a significant number of publications, notably investigating the transformation of monocots. Why monocots? For decades, crown gall appeared to be a disease of dicotyledonous plants alone, and monocotyledonous species were refractory to *Agrobacterium*. Indeed, crown gall tumors had not been observed in monocots in nature, and it was assumed that, somehow, *A. tumefaciens* was unable to infect them and transfer its T-DNA to their cells. This belief was held until the late 1980s, in spite of the fact that several articles had reported the detection of T-DNA gene products (such as octopine synthase) in monocot tissues cocultivated with *Agrobacterium* (Van Slogteren et al. 1984; Hernalsteens et al. 1984; Graves and Goldman 1986, 1987). Nonetheless, it was thought that only direct gene transfer would be able to produce transgenic monocots such as corn, rice, barley, and wheat. We now know, however, that these four plants in particular are readily transformable by *Agrobacterium* without the need for any special tricks (see for example Hiei et al. 1994), but this was not known a decade ago. Thus, protoplast electroporation appeared to be a viable alternative for transforming monocots. Unfortunately, monocot protoplasts tend to be very refractory to regeneration so that protoplast electroporation was not a good answer to the problem. Also, another difficulty with electroporation at that time was that it was thought (erroneously, as will be seen later) to work with protoplasts only; the plant cell wall was considered to be impermeable to DNA.

Another technique was thus needed. By now, the field of plant transgenesis had started to drift away from basic science to become resolutely mission oriented and patent driven, as evidenced by the increasing number of publications with authors who were employed by biotechnology companies. Also, the hunt for new DNA transfer technologies was beginning to be driven by proprietary rights placed on previously discovered ones. A fresh approach to monocot transformation was needed.

Thus, a methodology (that some claim could have been invented only in the United States) was born in 1987. It involved the use of modified firearms aimed at whole cells, which can be regenerated into plants much more easily than protoplasts. This technique became known as biolistics, particle bombardment, or the gene gun. It consisted in shooting (in the beginning, with a contraption housing an actual handgun) microscopic, DNA-coated metal (gold or tungsten) beads at a plant cell target, which could be, for example, a piece of callus or a leaf. The reasoning was that the velocity imparted to the beads by the firing of a gun cartridge would allow them to penetrate cell walls and deliver the DNA intracellularly. And it worked (Klein et al. 1987, 1988)! Now, there were some problems: Because of the very small size of the beads, aerobraking slowed them down and prevented cell penetration. Thus, the shooting had to take place in a partial vacuum. Next, the velocity of the beads had to be just right: Too much power and the plant cell sample was splattered all over the shooting chamber. This was difficult to achieve with regular cordite, so techniques were designed that allowed particle acceleration through the quick release of helium gas under pressure or an electric discharge. (The more advanced methods also did not require that the air be cleared of gun smoke after each experiment.) Much progress has been made in the simplification of the equipment, and there now exists a portable and inexpensive unit that has been used, notably, to transform rice by bombardment of seed-derived explants (Sudhakar et al. 1998).

Another justification to pursue DNA-mediated gene transfer experiments and develop new methodologies was the belief that many legumes would also remain refractory to *Agrobacterium*-mediated transformation. Legumes are of course dicots and do develop crown gall. The problem here is the existence of many cultivars with widely varying sensitivities to *Agrobacterium* strains and generally poor response to tissue culture conditions. In fact, the first genetically engineered legume, soybean, was obtained by biolistic transformation (McCabe et al. 1988). Here, again, a direct gene transfer approach seemed justified, one that would not necessarily rely on the patented gene gun technology.

It was demonstrated in 1990 that DNA could be transferred by electroporation to plant tissues and not solely to protoplasts (Dekeyzer et al. 1990). That, indeed, the plant cell wall did not represent an insurmountable barrier to the passage of DNA was subsequently observed repeatedly (see, e.g., Sabri et al. 1996). Thus, tissue electroporation was applied with some success to legumes, such as cowpea and bean via treatment of embryos (Akella and Lurquin 1993; Dillen et al. 1995) and pea meristematic tissues *in planta* (Chowrira et al. 1996; Chowrira et al. 1998). It is uncertain, however, whether tissue electroporation will prevail in the case of legumes, as various laboratories are actively screening many genotypes for sensitivity to *Agrobacterium* and have reported successful engineering in pea and chickpea, with a minimal amount of tissue culturing (for review, see Lurquin et al. in press).

Visits to laboratories involved in plant transformation show that many a gene gun and much electroporation equipment is now gathering dust on shelves. It is unclear what the future of direct gene transfer in general will be, as *Agrobacterium*-mediated gene transfer has become an extremely versatile technique in both dicots and monocots. Direct gene transfer has few advantages over the *Agrobacterium* approach, one of which is that the transgene of choice does not have to be recloned between T-DNA borders and introduced into *Agrobacterium*. However, very sophisticated *Agrobacterium* binary vectors containing multiple cloning sites, and screenable and selectable markers have been designed and do not really add any difficulties to the transgene cloning process. Further, *Agrobacterium* can be quickly and efficiently transformed by electroporation. In addition, direct gene transfer often (more often than *Agrobacterium*-mediated gene transfer) results in complicated foreign DNA integration patterns and complex segregation ratios in the progeny, or in gene silencing. On the other hand, *Agrobacterium*-mediated gene transfer does not allow mitochondrial and chloroplast transformation (although such a claim was made by De Block et al. 1985), because the

T-DNA is driven to the nucleus by the nuclear localization signals present in virD2. At the present time, there is no evidence for mitochondrial transformation in plants, but the chloroplast genome has been engineered by direct gene transfer, using PEG-mediated DNA uptake or biolistic transfer. Unlike the random foreign DNA integration observed in nuclear transformation, plastid transformation occurs by homologous recombination (Golds et al. 1993; Svab and Maliga 1993), which for years has been, and still is, the Holy Grail of nuclear transformation (see chapter 5).

Conclusions

There now exists a plethora of methods for the direct transfer of DNA into plant cells, and not a single one shown to work relies on the simple imbibition of dry seeds with DNA solutions, as described in chapters 1 and 2. It is in fact now known that passive uptake of foreign DNA by diffusion is an unlikely occurrence. However, if one takes into account the very rarely used techniques of protoplast microinjection (it is tedious and expensive) (Crossway et al. 1986); fusion of bacterial spheroplasts (*E. coli* or *Agrobacterium*) with plant protoplasts (which does not work any better than straight cocultivation of *Agrobacterium* with plant tissues) (Hasezawa et al. 1981; Hain et al. 1984); the silicon carbide fiber method, which injures the plant cell wall to facilitate DNA uptake (Songstad et al. 1995); and the microlaser technique to punch holes in cell walls (Weber et al. 1988), there are nine direct DNA transfer methodologies available, and many variants within each, that have been more or less successful. Most of these techniques have not survived the test of time, but I am not suggesting that their development was a waste of time. On the contrary, they were necessary, first to ascertain the reality of direct gene transfer, and then to optimize it. In that sense, the DNA-mediated transformation of plants is a great success. A look at the timetable of discovery shows that it took a mere three years (1984 to 1987) to develop the major methodologies still in use today. This is research proceeding at a very fast pace, especially if one considers that it took 17 years (1967 to 1984) to go from the first (unsubstantiated) claims of direct DNA transfer to the solution of the problem. Undoubtedly, this is a result of the quality of the later claims and the great technical advances in molecular biology in general. Without the advent of gene cloning, plant transformation would have been impossible. However, classical techniques such as sexual crosses also played a critical role in the analysis and confirmation of transformation events.

We saw in chapter 2 that unusual segregation ratios in *all* genetic crosses involving putatively transformed *Arabidopsis* and *Petunia* should have warned investigators that something was amiss with their material. Further, it should be remembered that these authors claimed that their putative transgenic primary transformants were homozygous for the added trait. The creators of the first real transgenic plants obtained by direct gene transfer were well aware of the crucial importance of genetic crosses in these matters and conducted a rigorous classical genetic analysis of the transformed individuals (Potrykus et al. 1985). Parental transgenic plants were found to be hemizygous (heterozygous), not homozygous, for the added trait (*neo*), which behaved in self- and backcrosses as a single, dominant Mendelian allele. In fact, homozygosity has not, to my knowledge, been reported in primary transformants. Further, there was tight correlation between the kanamycin-resistant phenotype and the physical presence of a single copy of the *neo* gene as detected by Southern blotting. Also, incomplete copies of the *neo* gene as well as copies of the pUC8 vector were physically linked with intact copies of the transgene, whereas kanamycin-sensitive offspring did not show the presence of foreign DNA sequences. This indicated the absence of segregation of donor DNA sequences and hence tight genetic linkage as well. The same results also confirmed foreign DNA tandem duplication and scrambling, as seen in pTi DNA experiments. However, some of the crosses resulted in non-Mendelian ratios, which were attributed to occasional transgene loss, possibly as a result of mispairing (or absence of pairing) of the transgene during meiosis in a hemizygous individual. Such phenomena have now been observed repeatedly, even in the case of transgenic plants obtained by *Agrobacterium*-mediated gene transfer (see chapter 3).

Will direct gene transfer into plants remain a viable technique? At this point, this field should not be considered basic research anymore. Rather, the issue is now to introduce genes of agronomic interest into plants in the most economical fashion and in a manner that does not infringe the patents of other companies, to maximize returns on research and development investments. In that regard, it is ironic but not surprising that both the particle bombardment technique and the whole-tissue electroporation technique have been patented, recently in the second case, by a company that did not even exist when the phenomenon was discovered (Dev and Hayakawa 1999). This state of affairs may motivate the search for alternative methodologies. Direct gene transfer has definitely joined the ranks of applied biotechnology, but can it truly compete with *Agrobacterium*-mediated gene transfer? Academic laboratories interested not in the marketing of transgenic plants but in using them as research tools to solve basic biological problems seem to have largely opted for *Agrobacterium*.

In the end, direct gene transfer into plants, especially using protoplasts as a model system, proved no different from DNA-mediated transformation in mammalian cells and even bacteria. All that was needed was a way to promote DNA entry into the cells, expressible cloned selectable markers, and a powerful selection protocol. Of course, this short and blunt statement does not do justice to the time and effort spent by dozens of individuals on this deceptively simple matter.

CHAPTER 5

Where We Are Now, and the Future

Genetically modified crops are presented as an essentially straightforward development that will increase yields through techniques which merely extend traditional methods of plant breeding. I am afraid I cannot accept this.... I believe that this kind of modification takes mankind into realms that belong to God, and to God alone.

PRINCE CHARLES OF WALES
From "Seeds of Disaster," *Daily Telegraph,* June 10, 1998

The Prince's extremely naive analysis fails to address the real problems of the planet. It reminds scientists of the remarks of another privileged individual. Queen Marie-Antoinette observed that if the hungry people in France were starving because they had no bread, then "Let them eat cake."

From "Let Them Eat Cake: Prince of Wales vs.
Biotechnology," press release from the Ninth
IAPTC Congress on Plant Biotechnology and
in Vitro Biology in the 21st Century.

Plant biotechnologists and Prince Charles are in clear disagreement. But unlike the downtrodden *sansculottes* who were soon to send Marie-Antoinette and her husband to the guillotine, much of the European public is now siding with a Royal Highness (or the Royal Highness is siding with the public). What is happening? Or rather, why are scientists so angry at an aristocrat not known for his penetrating views on scientific or societal problems? Also, why is it that the American public is so much more tolerant of products derived from genetically modified plants, which in the United States are not even required (with some exceptions) to be labeled as such?

In the next section, I will concentrate on aspects that I perceive as negative: potential ecological effects and socioeconomic impact. The following two sections, in contrast, aim to show that bright positives exist as well. This ambivalent attitude is probably shared by the public, academic and corporate scientists alike. Biotechnology raises entirely new ethical problems, with emotionally powerful points on both sides. The agrobiotechnology industry is now acutely aware of this societal phenomenon and will react to the challenges. As for myself, I am not pleading innocent; I have been involved in plant transgenesis research practically from the beginning and my laboratory is responsible for the recent generation of transgenic pea plants resistant to viral infection (Chowrira et al. 1998).

Plant Transgenesis and the Consumer

One of the promises of plant biotechnology at large is to provide safe, nutritious food for all, the premise being that the world will soon experience a shortage of agricultural products to feed a rapidly expanding population. It is, however, well known that the specter of overpopulation largely concerns the poorest continents (Livi-Bacci 1992). Thus, among other things, biotechnology promises a second "green revolution," which, like the first one, will be driven by discoveries made in industrialized nations, supposedly for the benefit of all. For example, a TV commercial funded by biotechnology companies and airing in the United States makes it clear that African farmers are included among the future beneficiaries of the new technology. (A later version of this commercial features the same actors but no longer contains any reference to Africa.) Many, however, think that the problem is not one of production; rather, according to them, the issue is one of equitable distribution within and between nations. In a scenario that shows the richest segment of the population declining in numbers and the poorest increasing, the problem becomes also one of just sharing of commodities. There is nothing biotechnology can do about that.

Even considering a population explosion taking place in the next few decades, the notion of freedom from starvation unfortunately now clashes directly with freedom to make a buck, which after all is what biotech companies are set up to do. Indeed, it should be remembered that the first green revolution was organized and financed by supranational organizations that distributed the products of their research and development, basically free of charge, to those in need. Plant varieties with better yield and particular climate adaptations were not patented. This picture has changed drastically in the last decade or so, with the patent craze now gripping companies and academia alike. Not only are transformation techniques patented, so are commercially relevant transgenes and transgenic organisms. It is expected that a second green revolution would benefit developing countries, much as the first one did. However, these countries would no longer be the beneficiaries of research and development done elsewhere or even at home (China and India, for example, have extensive plant biotechnology programs), as they would be at the mercy of giant industrial conglomerates holding all the patents.

This is best illustrated by the infamous "terminator" technology scenario attributed to Monsanto (but denied by the company). In this scheme, a "terminator" gene in high-quality (through genetic engineering) cereals, for example, would be triggered before harvest and make all seeds sterile. This would prevent farmers from keeping part of their harvest for replanting at a later date. Conclusion: Farmers become entirely dependent for their livelihood on the company that developed that crop, or they decide to grow a nonpatented variety of lower quality. This is no way to teach self-reliance to the developing world, a pet project entertained by developed nations. Finally, many crops such as millet, sorghum, and cassava, traditionally cultivated in less-developed nations, are almost completely ignored by biotechnology companies; they are considered small crops and are not much grown in industrialized countries anyway. As one corporate senior scientist once said, "Biotechnology companies are not the Rockefeller Foundation."

The proponents of a second green revolution see the first one as a great success. This is true from the global perspective of total food production. However, this view ignores the fact that the green revolution has made millions of poor farmers even poorer while the rich grew richer. Indeed, the new crop varieties distributed in developing nations required fertilizers and pesticides that small farmers could not afford (fig. 5.1). These people were forced to sell their land to rich farmers with whom they could no longer compete. As a result, these landless farmers experienced mass exodus to already overcrowded cities in which they were forced to live in slums. The net result was an increase in poverty, not the opposite (Lappe et al. 1979).

The multifaceted aspects of green revolutions have been recently reviewed by an official of the World Bank (Serageldin 1999). In this article, I. Serageldin stresses that the private sector is unlikely to invest in the needs of developing countries, given that monetary returns would be few. He advocates the creation of deeper partnerships between biotechnology companies and public sector research institutions in developing countries (such as those already established between Monsanto and Kenyan and Mexican

FIGURE 5.1.
A barley field in the village of Phek, central Nepal, in 1979, sown with the first green revolution seeds. The farmer was unable to afford the necessary agrochemicals, hence the scarcity of growth and poor harvest. It can validly be argued that this farmer would benefit from pest-resistant, fertilizer-independent, genetically engineered seeds. (Source: Linda Stone, with permission.)

research institutes), while preserving intellectual property rights and royalties in targeted markets. In conclusion, a second green revolution, based on biotechnology and aimed at improving human nutrition in the developing world, should be preceded by a careful analysis of the benefits and pitfalls of the first one.

Nevertheless, a second green revolution has started in the Western world. Vast acreages in the United States and Canada are planted with genetically engineered canola, cotton, corn, and soybean. These plants have been engineered with either herbicide- or insect-resistance genes. Two multinational corporations have been at the source of genetically engineered herbicide resistance in plants: Monsanto and AgrEvo. It will be recalled that Monsanto was, starting in the early 1980s, at the forefront of plant biotechnology. Much basic science was done there under the leadership of, among others, Ernie Jaworski, Robb Fraley, Bob Horsch, and Steve Rogers. This research culminated in the development of several generations of *Agrobacterium* vectors containing one or several genes coding for the detoxification of the Monsanto broad-spectrum herbicide, Roundup (glyphosate), which inhibits the synthesis of the aromatic amino acids tryptophan, tyrosine, and phenylalanine and hence causes plant death (Hinchee et al. 1988). The research that led to plants engineered with a gene coding for the resistance to Basta (also known as phosphinothricin, glufosinate, PPT, and bialaphos, another nonselective herbicide) was different, in the sense that the company (AgrEvo) now producing PPT was not involved in the discovery, cloning, sequencing, and expression of the resistance gene. Rather, the basics were accomplished by a collaboration between two small companies, Plant Genetic Systems (PGS) in Belgium, and Biogen in Switzerland (Thompson et al. 1987; De Block et al. 1987). PGS itself was a commercial offshoot of the University of Ghent laboratory (originally under the direction of Schell and Van Montagu), where so much about the crown gall system was discovered. PGS has been acquired recently by AgrEvo, which now not only produces the PPT herbicide but also engineers crop plants with the resistance gene, *bar* (or *pat*), which detoxifies PPT by acetylating it (PPT inhibits the enzyme glutamine synthase, which plays a central role in the metabolism of ammonia).

The moral of the story is the same in both cases: The companies manufacturing the herbicides provide growers with the seeds of the resistant transgenic plants plus, of course, the relevant herbicide. This makes a lot of sense from a business point of view. But does this approach benefit anyone other than the companies themselves? In North America, where agriculture is extensive, reliance on a single herbicide with good environmental and toxicological properties, that kills all competing plants and leaves the engineered crop plants unscathed, is certainly a welcome simplification. However, it is hard to see how this would benefit the small farmer in Africa, Asia, or even Europe.

Another drawback is the potential narrowing of the gene pool. If indeed only certain herbicide-resistant cultivars remained available, what would happen if an unforeseen epidemic were to strike these plants? This has happened before, with catastrophic consequences, and could happen again. Another unknown is, of course, the appearance of weeds resistant to these herbicides, either by spontaneous mutation and selection, or by gene transfer. One could argue that in that case, crop plants would then be engineered with yet another herbicide resistance. However, the current problem of human pathogen resistance to antibiotics uncomfortably comes to mind and parallels can be seen immediately.

Another argument used by genetic engineering companies is that plants engineered with herbicide-resistance genes will ultimately result in less waste and higher productivity. This is certainly true on paper; with fewer competing weeds, productivity will increase. Now, assuming that growers pay the same price as before for engineered seeds and use a single herbicide instead of a cocktail (less herbicide use is also a promise made by the biotechnology industry), given higher productivity, costs of production per acre or bushel go down and the farmer's profit margin gets bigger. In a competitive market, where everybody uses the same seeds and the same herbicide, this should lead to a *decrease* of the cost to the customer, the farmer, and eventually the consumer. It is doubtful that this has already happened, but it remains an interesting test of the promises of biotechnology. After all, if consumers pay the same or more for products that contain a foreign gene of absolutely no interest to them, what is the point? The same reasoning should hold true for crop plants engineered with the *Bacillus thuringiensis* (Bt) toxin, which makes them resistant to certain insects.

Naturally, financial considerations do little or nothing to alleviate concerns about the effects of these foreign genes on human health and impact on the planet's ecology. So far, there is no evidence that crops engineered with the Bt toxin, PPT-resistance, and glyphosate-resistance genes have had any adverse effect on humans. Indeed, humans do not have the equivalent of insect intestinal cells, which are the targets for the Bt toxin, nor do they possess 5-enolpyruvylshikimic acid 3-phosphate synthase or glutamine synthase, which are inhibited by Roundup and Basta, respectively. This is not to say that other genes conferring resistance to herbicides or insects would be harmless; the metabolites synthesized by these gene products, which do accumulate in engineered plants, must be individually tested for toxicity. For example, engineering plants for resistance against the herbicide 2,4-D was once contemplated (Streber and Willmitzer 1989). This could be done by transforming plants with one or several catabolic genes (Perkins and Lurquin 1988; Perkins et al. 1990) from the soil bacterium *Alcaligenes eutrophus*, which has the ability to degrade 2,4-D and, in fact, uses it as a carbon source. The problem is that one of the degradation products in the catabolic pathway is 3,5-dichlorophenol, a suspected human carcinogen. Clearly, such plants would not be popular with the consumer, were the consumer to be enlightened.

In terms of ecological impact, the use of the Bt toxin gene, for example, will undoubtedly select for mutant, resistant alleles in insect populations. The assumption made by scientists was that such mutations would be recessive, and would thus only extremely rarely lead to the appearance of resistant insects, which would have to be homozygous for the resistant trait. In truth, transgenic plants harboring the Bt toxin gene would not impact the environment any more than the older practice of spraying the toxin itself.

However, a recent report has shown that insect resistance to Bt toxin may not be recessive after all; it seems to be semidominant, at least under laboratory conditions (Huang et al. 1999). Such an observation should lead to a reconsideration of the use of the Bt toxin, either applied exogenously or produced endogenously by transgenic plants. On the other hand, there are many naturally occurring Bt toxins, and it is quite conceivable that if resistance to one variety were to occur on a large scale, biotechnologists could easily reengineer plants with other Bt toxin genes for which no resistance yet exists. Further, there is as yet no evidence that transgenes coding for herbicide resistance have been introduced on a significant scale by cross-fertilization or other natural means to weed species.

The issue of human health brings up the question of labeling and identifying genetically engineered products and their derivatives. In the United States, federal legislation has made this step unnecessary (except in matters of potential changes in allergenicity), and hence the public usually has no way to know if the cereal boxes and produce shelves contain genetically modified foodstuffs. It is interesting to note that the first genetically modified plant product, the FlavrSavr tomato, was a big market flop. Apparently, this tomato was not much better than its regular competitors, but also it was labeled as a genetically engineered product and did not gain the public's sympathy at all. Some restaurants in New York City, in particular, even proudly posted that no genetically engineered vegetables were on the menu. The new law has made such protests much more difficult, if not impossible, and the American public has been largely silent ever since.

The situation in several European countries is different, to such an extent that opponents of genetically engineered foods have taken on the appearance of a new Luddite movement. Tomato paste made with genetically modified tomatoes had to be taken off the shelves in Great Britain, and in Switzerland a referendum to ban any type of genetic engineering was supported by many citizens (but ultimately defeated at the ballot box). The situation in France and Germany is not much better. People do not like the idea of eating anything whose genome has been tampered with. In some cases, trial plots of engineered plants have been savaged. It is difficult to know exactly why Europeans are so adamant about this issue. Some have invoked the specter of Nazism and its eugenic policies, but it is hard to believe that the abhorrence of such policies would extend to plants. Further, Germany, as well as many other European countries, has thriving genetic research programs unrelated to biotechnology. In my opinion, this negative reaction is the result of a much greater political awareness and the generally poor opinion that Europeans have of multinational corporations and big business, especially of U.S. origin. Also, European cultures have never embraced the United States' "can-do" attitude and its corollary, that "new and improved" is necessarily better. A study conducted in 1996 and 1997 and published in 1999 (Gaskell et al. 1999) suggests that at least three factors distinguish Europeans from Americans in their perception of genetically modified foods. First, the European press has provided since 1993 significantly greater coverage of biotechnology issues than its U.S. counterpart. Interestingly, press coverage in Europe tends to be more positive than in the United States, suggesting that the more exposure a controversial issue gets, the more negative perception it gains. Next, this study shows that trust in governmental authorities who decide on food safety regulations is higher in the United States than in Europe. This observation fits well with the very cynical attitude that many Europeans hold toward their elected representatives and governmental bureaucracies. Finally, even though knowledge of simple biotechnology concepts is greater in Europe, false and threatening images of the risks of biotechnology are there regarded as true much more frequently than in the United States. One issue not addressed by the authors of this study (who are British) is that of European versus American cuisine. Although Great Britain does not have an impressive culinary tradition, this is not so for many other European countries. I propose that for people belonging to a culture in which food preparation and consumption are integral parts of the cultural heritage, tampering with meals is a violation much greater than the principle of biotechnology itself. This would not apply to the largest part of U.S. society, whose view of food is much more utilitarian and expedient (simply refer to the "deflavorizing machine" mentioned by Woody Allen in his film *Broadway Danny Rose*). Thus, I believe that Europeans will have to be convinced that spaghetti made with herbicide-resistant wheat, cassoulet prepared with genetically modified beans, and paella concocted with fungus-resistant rice, not to mention Beaujolais or Valpolicella based on engineered grapes, will taste the same as before.

In brief, the issue of genetically modified plants (and, by extension, all organisms) seems to hinge on two factors: First, fortunes will be made if genetically modified organisms are accepted by the public as good and necessary; second, the public is largely uninformed about the risks and benefits of genetic engineering in agriculture and has probably reacted too emotionally. As of now, these two factors are incompatible and irreconcilable, and what is needed is better communication between scientists (preferably not those working for biotechnology companies) and the consumers. The worst possible way to handle the problem, however, is to pass legislation sweeping the issue of genetically modified organisms under the rug.

As for Prince Charles versus biotechnology, it is understandable that well-meaning scientists were outraged to see their turf invaded and years of work summarily dismissed. However, his newspaper article raised reasonable questions that the scientists did not address in their own press release. This short exchange is possibly a good portrait of the general problem: absence of communication.

Biotechnology companies in the United States have finally understood this problem. A campaign of TV commercials, featuring biotechnologists and multiethnic beneficiaries of the new technology, began in April 2000. Needless to say, only beautiful, happy people are shown, and among them are benevolent scientists and overjoyed farming families. No effort was made to make these commercials actually informative, other than presenting biotechnology as good for you. It remains to be seen whether a TV-commercial-saturated public will be convinced by these actor portrayals and as yet unsubstantiated promises.

Last, but not least, one may wonder whether transgenic plants resistant to bacterial, fungal, viral, herbicide, and insect attacks are really necessary to maintain grain production per capita constant. Experts disagree, some arguing that present food production has already outpaced population growth (Serageldin 1999). Then, in whose interest is it to develop these creatures? In the not so old days, transgenic techniques were seen by academic scientists as a great tool to understand how plants worked. How things have changed! Fortunately, not all transgenic research is aimed at maximizing the profits of biotechnology and seed companies, with their emphasis on plant protection. Recently, academic laboratories have embarked on research destined to enhance the nutritional values of plants by manipulating, among others, levels of provitamin A ("golden rice") and vitamin E, as well as micronutrient (such as iron) content (DellaPenna 1999). Hopefully, the distribution of such plants, if they are ever produced commercially, will be largely independent from big business.

Finally, it should be noted that the corporate world of agrobiotechnology may be evolving fast. A Monsanto–University of Washington team announced on 3 April 2000 that they had obtained a rough draft of the rice genome, covering about 80 percent of its length. This achievement comes years earlier than anticipated by competitors. Even more surprising was the announcement that the data would be made available to the public. This has never been done before by the industry. Is it an indication of change?

In this section, I have expressed concerns about the applications of the science of plant transgenesis and their potential impact on the environment, social justice, and human health. These concerns do not represent a condemnation of plant biotechnology as a whole. Rather, they call for a hard evaluation of the use of applied, corporate biological science in human affairs. After all, nobody has ever questioned the importance of ethics in the genetic manipulation of human beings. Why should there be lower (or no) standards in the case of our most basic needs: crop plants?

Frontiers in Transgenesis

Quite certainly, not all research with transgenic plants is of an applied or industrial nature. Transgenesis has become an important tool in the understanding of basic plant biology problems such as development, differentiation, gene regulation, and signal transduction. It is now common practice to complement newly discovered mutations by introducing wild-type transgenes to ascertain the exact nature of the mutation and its phenotypic effect(s). Further, gene regulation can now be studied by cloning the coding sequences of reporter genes under the control of a variety of promoters responding to, for example, light, temperature, or other external and internal signals. It is impossible to do justice to all the basic work involving transgenesis that is taking place today. Therefore, only a few examples will be developed here to illustrate where and why progress is needed.

Given the present and future whole-genome sequencing projects with *Arabidopsis* (almost complete at the time of this writing) and rice, and the recent availability of relatively inexpensive microarrays, it will be crucial to identify the function of many heretofore undiscovered genes. This gene discovery process will necessitate a highly selective method of gene disruption (or "knock-out") and replacement, which, unfortunately, does not yet exist in plant systems. Gene disruption has long been used in bacterial, fungal, and animal systems to attribute phenotypic effect(s) to a DNA segment of unknown function. The same technique can work in reverse, and it is then sometimes known as gene replacement: A mutant gene can be specifically replaced by a correct copy, or vice versa, at the original locus. This approach is being contemplated for gene therapy in humans. Gene disruption/replacement is based on homologous DNA recombination involving a double crossover between the genomic target gene and the incoming, cloned disrupter/replacer gene. The phenomenon is also called gene targeting because it allows the integration of a transgene at a precise target in the chromosomal DNA. Unfortunately, gene targeting works very inefficiently in plants, and nobody has so far figured out how to select for such events.

In yeast (but not in mammalian cells), integration of a foreign gene occurs at much higher frequency if the transgene is flanked by a region homologous to chromosomal DNA. This type of enhanced insertion frequency was not observed in direct gene transfer to tobacco protoplasts (Lurquin and Paszty 1988), and taken together with the largely negative results of a number of well-designed experiments, it argues that gene targeting in plants must be a very rare event. Homologous recombination of a transgene has, however, been demonstrated by genetic complementation between two truncated versions of the *neo* gene (Paszkowski et al. 1988). For this, tobacco protoplasts were transformed with a vector containing an end-deleted, inactive version of the *neo* gene. Plants were then regenerated from these protoplasts. Fresh protoplasts were in turn prepared from these transgenic plants and retransformed with another truncated version of the *neo* gene, deleted at the other end, the two versions sharing a roughly 500-base-pair-long region of homology. Homologous recombination events between the two truncated *neo* genes were selected for by plating the retransformed protoplasts in the presence of kanamycin. Kanamycin-resistant clones appeared at a frequency of 0.5 to 4.2×10^{-4} relative to transformation with an intact transgene. In other words, gene targeting by homologous recombination occurred at an absolute frequency of 10^{-7} or less. This number is higher when homologous recombination experiments are performed with *A. tumefaciens* vectors (because *A. tumefaciens*–mediated transformation of protoplasts is more efficient), but the efficiency of targeting itself remains low, even in this system. The same situation is encountered with mammalian cells. Of course, the experiment described here was used as a model to demonstrate the feasibility of gene targeting in plants, and it worked because the targeted chromosomal locus was an inactive version of a selectable marker that could be selected for only if this gene could be reconstructed by single homologous recombination. This is very different, however, from attempting to knock out or replace an anonymous, nonselectable, specific chromosomal locus by double crossing over.

Capecchi's group designed an elegant, and now classical, positive/negative selection system that allows detection with much greater efficiency the rare recombination events leading to genuine gene replacement by double crossing over in mammalian cells (Mansour et al. 1988). Basically, the gene to knock out, which must first be cloned, even as an anonymous locus, must be interrupted by a *positively* selectable marker, such as *neo*, by *in vitro* manipulation. In addition, the cloned gene of interest must be flanked by chromosomal DNA normally surrounding it *in vivo*. Next, this construct is further manipulated through the addition of a *negatively* selectable marker, such as the thymidine kinase gene of the herpes simplex virus (HSV-*tk*). The product of HSV-*tk*, unlike the endogenous thymidine kinase, has the ability to phosphorylate the base analog ganciclovir, which, when incorporated into DNA, blocks further DNA elongation. In other words, a cell transformed with both the *neo* and HSV-*tk* genes will survive when exposed to G-418 (used instead of kanamycin in mammalian cells) but will die if ganciclovir is also present in the medium. A cell that expresses *neo*, but *not* HSV-*tk*, will survive in the presence of both G-418 *and* ganciclovir.

Figure 5.2 describes how the technique works. If cells transformed with a construct containing a chromosomal locus interrupted by the *neo* gene and harboring the HSV-*tk* gene integrate the recombinant plasmid randomly, both the *neo* and the HSV-*tk* genes will be incorporated and expressed in the transgenic cell lines. Thus, these lines will become resistant to G-418 but sensitive to ganciclovir, and they will not survive in the presence of both compounds. If, on the other hand, integration within the host genome occurs by double crossover between the regions flanking the *neo*-interrupted locus and the homologous regions in the host's chromosomal DNA, only the *neo* gene will become integrated, whereas the segment of vector carrying the HSV-*tk* gene will be lost. Such lines will become resistant to both G-418 and ganciclovir. In other words, this methodology distinguishes between random integration and integration by double homologous recombination— that is, gene targeting and replacement. In the example given, a locus was interrupted by the *neo* gene and its function was thus impaired. The gene so disrupted, when it replaces its wild-type counterpart in transgenic lines, may cause the appearance of one or several phenotypic effects (or not) and thus assign (or not) a phenotype to that particular gene. The functions of many genes have been determined in this way. It should, however, be noted that this technique does not allow for the selection of single recombination events.

Does this technique work in plants with these selectable markers? Unfortunately, the answer is no. Ganciclovir and other similar analogs do not function as good negative selectable marker in plant systems. A few negative selectable markers adapted to plant cells have been reported (Depicker et al. 1988; Nussaume et al. 1991; Stougaard 1993), but, to my knowledge, they have not been used in conjunction with positive selectable markers to investigate homologous recombination. Thus, gene disruption and genotype/phenotype assignments cannot yet be done using this powerful methodology. This is even more unfortunate when one considers that whole plant genome sequences will be available soon, without the existence of an adequate technique to study the thousands of new genes that will be discovered. Further, gene targeting has practical implications for the

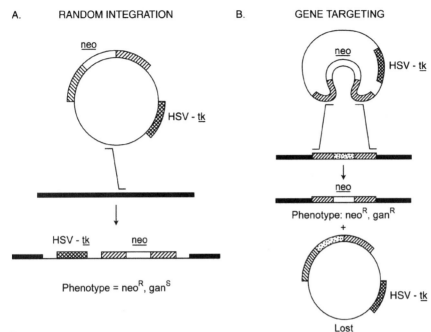

A.   RANDOM INTEGRATION

B.   GENE TARGETING

Phenotype = neo^R, gan^S

Phenotype: neo^R, gan^R

Lost

FIGURE 5.2.
Strategy used to select for double-recombination events (gene targeting) involving nonselectable chromosomal genes in animal cells. A: Random chromosomal integration via single crossover of the plasmid vector harboring *neo* (flanked by host chromosome sequences) and HSV-*tk*. Both *neo* and HSV-*tk* end up integrated at the same locus. These transfectants become resistant to kanamycin but sensitive to ganciclovir. The original chromosomal locus is inactivated. B: Targeted chromosomal integration via homologous sequences flanking *neo*. In this case, only *neo* becomes integrated via double homologous recombination between vector and host sequences. The HSV-*tk* gene is eliminated, together with vector sequences and the original chromosomal locus at which the double crossover took place. The eliminated sequences may be circularized or not, but the HSV-*tk* gene will eventually be lost (degraded), because its vector lacks an origin of replication. The phenotype of these transfectants will thus be kanamycin resistant and ganciclovir resistant, the latter because of the loss of the HSV-*tk* gene. (Adapted from Mansour, S. L., K. R. Thomas, and M. R. Capecchi. 1988. Disruption of the proto-oncogene *int-2* in mouse embryo-derived stem cells: A general strategy for targeting mutations to nonselectable genes. *Nature* 336:348–352.)

genetic engineering of crop plants. Many are concerned that random integration of transgenes may have cryptic effects on the host plants—that is, undesirable effects with no immediate phenotypic impact but with potentially harmful consequences for the consumers of this plant material. This could happen, for example, if the synthetic pathways of secondary metabolites were inadvertently disrupted by transgene insertion. Toxic compounds might then accumulate in the host. This problem could be avoided if the transgene could be targeted to, for instance, a host pseudogene or an inactive transposon.

Other methods of gene "tagging" do exist in plants, such as transposon (Grandbastien et al. 1989) and T-DNA insertion or Cre-*loxP*-mediated integration (Odell et al. 1990), but these events are random (even integration of *loxP* cannot be controlled at the present time) and thus offer a far less powerful approach than *bona fide* gene targeting. Granted, yet other techniques of gene disruption exist or are being developed in plant systems (Sommerville and Sommerville 1999). Also, there currently exist large collections of insertion mutants that will be useful in assigning functions to newly discovered genes (Sommerville and Sommerville 1999). However, these techniques do not presently allow gene replacement or targeting by homologous recombination. Perhaps this problem will soon be solved, and I do hope that the elegant Capecchi model remains a source of inspiration for plant biologists.

Another promising area of research is that of metabolic engineering, in which whole metabolic pathways can be dissected and potentially reengineered through transgenic techniques. This will not be easy to do, however. Present-day organisms have been fine-tuned by evolution, and their metabolic pathways are highly integrated. Nevertheless, scientists at DuPont and elsewhere have engineered corn and soybean lines with enhanced amino acid contents as well as altered compositions of fatty acids and seed storage proteins (Mazur et al. 1999). Further, it is not yet possible to transfer very large pieces of autonomously replicating DNA to plant cells, because the equivalents of bacterial and yeast artificial chromosomes have not yet been developed. Such tools would facilitate the study of large gene clusters whose expression would need to be coordinated.

An interesting spin-off of transgenic research was the rediscovery of gene silencing through duplication. This phenomenon was discovered decades ago and, in plants, sometimes called paramutation. Its physical basis was not understood. The rediscovery of the phenomenon, now called cosuppression, was made in the wake of attempts to inhibit gene expression using antisense technology. This technique consists of cloning the coding sequence of the endogenous gene whose expression is to be inhibited, in the reverse (antisense) orientation, under a promoter of choice (van der Krol et al. 1988). This is actually how the Flavr Savr tomato was created (Sheehy et al. 1988). Originally, it was thought that the antisense gene, producing mRNA molecules of polarity opposite to that of the endogenous mRNA and complementary to them, would lead to the formation of duplex mRNA untranslatable by the translation apparatus. In fact, these double-stranded mRNA molecules are not detectable (they may turn over extremely rapidly), and it is now thought that antisense effects occur at both the level of transcription and that of translation (Cornelissen and Vandewiele 1989). Further, antisense inhibition of endogenous genes is variable and not necessarily correlated with antisense gene copy number or even expression.

Unexpectedly, it was discovered by doing appropriate control experiments that the transgenes need not even be in the antisense orientation to exert their inhibitory effect. Transgenes in the sense orientation could do just as well! Thus, plant cells somehow have the ability to "measure" gene copy number, a property that can lead to gene silencing when the number of homologous genes exceeds ploidy level. Several hypotheses have been offered to explain this effect. These include, but are not restricted to, foreign DNA methylation, base-pairing in stem-loop structures formed by multiple copies of a transgene, and detection and degradation of mRNAs exceeding a certain concentration threshold (Matzke and Matzke 1995; Meyer and Saedler 1996; Depicker and Van Montagu 1997; Wolffe and Matzke 1999). Unfortunately for the genetic engineer, *more* in this case does not necessarily mean *better*; accumulating transgenes for a desirable effect, more often than not, results in the inhibition of genes coding for that trait! Of course, cosuppression also offers the possibility of inhibiting undesirable traits by transforming plants with transgenes homologous to those coding for that trait.

Recently, a broad and original model for cosuppression has been proposed (Jorgensen et al. 1998). This model rests on the observations that cosuppression can spread systemically throughout a whole plant, like a viral infection, and that, as is seen in these infections, cosuppression can produce mosaic effects in plant organs. Indeed, cosuppression of a gene involved in pigment synthesis can produce white (cosuppressed) spots in petals, much as dark green, uninfected spots can be detected on the leaves of plants infected with an RNA virus. These dark green islands do not contain viral RNA, strongly suggesting the existence of a specific RNA degradation system. Thus, cosuppression may involve a mobile signal molecule able to travel through the phloem and disperse into surrounding tissues and, by analogy with resistance to viral infection, this signal molecule may be RNA. These RNA molecules, according to the model, might be produced from an RNA template (for example, mRNA molecules transcribed from a transgene) by an endogenous RNA-dependent RNA polymerase. These cRNA molecules, conceivably as part of a ribonucleoprotein complex, could then be transported through the plant, pair up with their antiparallel mRNA transcripts, and lead to degradation of these duplex structures by a double-stranded specific ribonuclease, and, hence, lead to phenotype suppression. In that sense, cosuppression would be the result of an evolutionary response to RNA virus infection acting through an RNA-mediated signal transduction process over long distances. Significant experimental support has been provided in favor of this model; microinjection experiments have indeed shown that RNA (but not DNA) becomes associated with a specific protein and that this complex can be translocated via the phloem over long distances (Xonocostle-Cázares et al. 1999). If this model is further sustained by experimental evidence, an entirely new window will be open on the mechanism of gene regulation in plants. Interestingly, small antisense RNA molecules, approximately 25 nucleotides long, have been discovered in cosuppressed transgenic plants and plants infected with potato virus X, but not in control, uninfected, or nonsuppressed transgenic individuals (Hamilton and Baulcombe 1999).

Cosuppression may then indeed have evolved as a weapon against virus infection. Results from Vicki Vance's laboratory have shown that the 5'-proximal region of the tobacco etch potyviral (TEV) genome has the ability to suppress gene silencing. In the absence of this sequence, gene silencing is maintained in cosuppressed transgenic plants. This suggests that posttranscriptional gene silencing evolved as a defense mechanism aimed at limiting viral RNA accumulation in infected plants. The acquisition of the 5'-proximal region by TEV then allowed it to defeat this mechanism (Anandalakshmi et al. 1998). This pioneering work should allow fine-tuning of gene expression in plants containing multiple copies of a transgene.

The examples of gene knock-out and replacement (still to be developed) and cosuppression (with its reversal) should suffice to demonstrate that transgenesis has added and will add substantially to basic plant science. In fact, cosuppression, and its own suppression, would not have been amenable to molecular study without the existence of transgenic plants. Nevertheless, since plant biotechnology joined the realm of applied science years ago, it is inevitable that many more very practical applications will follow. Fortunately, many current projects go well beyond plant protection. For example, plants are now being engineered with human pathogen genes coding for proteins that trigger an immune response in infected individuals. The idea is to develop edible vaccines that would require no syringes for inoculation (and hence no sterilization equipment) and no refrigeration. Such vaccines would be ideal for use in developing countries where medical facilities are scarce. Also in the medical domain is the intriguing possibility of producing human blood substitutes in plants. It has been recently demonstrated that transgenic tobacco containing human α- and β-globin genes do synthesize hemoglobin Hb A (Dieryck et al. 1997). Whether such plants will one day replace blood banks is of course an open question. Finally, plants have been genetically modified with bacterial genes coding for the detoxification of organic and heavy metal pollutants, not only allowing them to survive in polluted areas but also making them able to sequester or degrade these pollutants. Here, also, large-scale applications have not yet started, but the promise of land reclamation is an attractive idea.

Conclusions

We have seen in this book that transgenesis in plants was initiated as a wild-goose chase in the late 1960s, at a time when crude molecular biological techniques could not have made such an ambitious project successful. In the early 1980s, the prospects of plant genetic engineering had become brighter: An issue of the *National Enquirer* even devoted a full-length article to the future of the field. In this piece (featuring a grainy photo of Edward Cocking that made him look like an escaped convict), cactus plants were engineered with genes that imparted to them legs, and thus mobility, plus appendages functioning as submachine guns shooting seeds. These plants could be used as bodyguards or private police. Also, oleaginous plants had been engineered to produce gasoline, and their new genes even allowed them to pump it directly into your car! I must add quickly that the journalists of the *National Enquirer* doubtless exaggerated what the scientists told them, but, on the other hand, who could have foreseen tobacco plants as human blood donors or bananas as vaccines! Similarly, it is not out of the question that the shortage of fossil fuels and their eventual disappearance may prompt scientists to engineer plants for higher methanol yields, for example. At roughly the same time the *National Enquirer* article came out, I was interviewed by a journalist working for a local eastern Washington newspaper, who wanted to know how farmers could possibly benefit from this emerging technology. Eager to use a Washington State example, I suggested that genetic engineering might one day be used to make strawberry plants resistant to frost. He found this all very funny and entitled his article "Scientist Plans to Grow Frozen Strawberries." I don't think such strawberries exist yet, but cold-tolerant plants have indeed been engineered, and even "smart" plants (not necessarily in a true intelligent sense) may one day appear. In brief, what seemed ludicrous yesterday is almost real today and will be mundane tomorrow. As for cacti taking the place of German shepherd dogs or armed humans. . . .

It has been a long time since the *National Enquirer* has published anything on genetically modified plants. But, generally speaking, only the most extreme applications of a technology will spur public interest. In the case of plant biotechnology, unfortunately, only the most controversial applications, such as those dealing with crops destined to be eaten by people, have been discussed or perhaps simply mentioned in open forums (e.g., the mass media). Plant biotechnology has, in the minds of many, been condemned before getting a fair hearing. Most, if not all, of the benefits that this technology has brought to basic science and less controversial applications have been ignored or misrepresented. The onus is now on scientists to communicate openly, and in a user-friendly manner, with a skeptical public to show that science can work for the people, not in spite of or against them.

Certainly, one of the more honorable goals of plant biotechnology would be to eradicate hunger. According to one interpretation of international statistics (Mann 1999a), increases in the world production of the three main crops, rice, wheat, and maize, have stalled. This, according to some, should ring alarm bells, because a slower increase in production rates will eventually mean a reduced availability per capita in a growing population. Agricultural biotechnology is of course seen as a problem-solver, although the challenge of manipulating polygenic quantitative traits such as yield is considerably beyond our present-day ability of adding one or two single genes. Other approaches, also based on genetic engineering, such as better control over gas and water exchanges between plants and the atmosphere, as well as improving photosynthetic yields, are being considered (Mann 1999b). These are all worthy basic scientific problems with great potential for field applications. But is the world truly facing a food crisis? Or are the Cassandras acting up? A graph published in *Science* (Mann 1999b) shows that the 1997 world grain production was about 353 kg per capita, well below the 1985 high of 372 kg per capita. However, the same graph shows wide fluctuations of this value over the past 31 years. For example, the numbers were 329 kg per capita in 1995 and 330 kg per capita in 1975—that is, the same value 20 years apart. The 1966 low was about 312 kg per capita. Linear regression analysis of these data shows a weak correlation coefficient (.56), which makes trend analysis difficult. Clearly, many factors influence grain production and availability. Even during the heyday of the green revolution, between 1965 and 1985, important fluctuations can be seen, which almost by definition should not have been present. The work of the Indian scholar Amartya Sen, 1999 Nobel laureate for Economic Sciences, for example, has shown that food production is only one element, and not necessarily the major one, in the complicated chain of events that can lead to disastrous famines. Other factors such as unemployment, speculative stockpiling, and fall in purchasing power have caused starvation in the Third World even in periods of high agricultural output. Nevertheless, modest attempts at bringing plant biotechnology to poor nations have started. Academic laboratories, often organized in international consortia and, in some cases, in collaboration with the private sector, have embarked on programs aimed at engineering oil palm, banana, cassava, and indigenous potatoes (Moffat 1999). These programs are in their infancy, and it remains to be seen whether local populations really need and accept these genetically modified crops. Also, it is unclear whether developing nations are currently facing a shortage of these plant products. In the case of bananas and oil palms, one has the dark suspicion that, again, it is the owners of the plantations (generally multinational corporations) that will benefit most, not the people doing the actual work of growing and harvesting these commodities, and not the consumer.

Notwithstanding these concerns, vigilance over food production is needed. On the other hand, "catastrophism" is unlikely to convince a jaded public that has already heard it all from scientists eager to secure lucrative grants. The history of science shows that, most often, applications following basic discoveries appear when they are needed; it is not the availability of the discovery that creates the need and drives the application. There is no reason to believe that plant biotechnology should behave otherwise, or that it should contribute to create needs that do not really exist. Absolute food production is not seen by all as a limiting factor in an expanding population. Many think that cultural, political, and economical objectives drive the food production process much more directly than human nutritional needs (Bodley 1996). Thus, the emotional arguments presented against biotechnology by Charles Windsor in his *Daily Telegraph* article may indeed sound naive, but the rebuttal offered by scientists may be equally so.

Finally, it can be said that much of the recent public outrage raised by the marketing of transgenic food plants is in part the result of the rather arrogant attitude initially exhibited by some major biotech companies. Had these companies invested time, capital, and effort into public education, it is quite possible that opposition to their transgenic products would have been much mitigated. In that context, the January 2000 Montreal accords regarding the international trade of genetically modified organisms and their labeling is a step in the right direction.

In the end, acceptance of transgenic foodstuffs will depend on stringent, science-based, safety tests accompanying the promise of enhanced quality. In that light, the actions of fringe elements of society, responsible for the destruction of test plots and laboratories where transgenic research is being conducted and genetically engineered foods tested, can never be condoned. These unfortunate events have taken place in both the United States and Europe. Recently, in the state of Washington, usually known for its progressive policies, strawberry plots and greenhouses belonging to Washington State University have been savaged, even though they contained not one single transgenic plant! In fact, nobody at that university has ever conducted applied transgenic research on strawberries. One is then left to wonder how the perpetrators selected their irrelevant target. It is unclear whether these acts of vandalism are orchestrated and financed by larger organizations and how their action plans are designed. What is clear, however, is that a movement such as Greenpeace caters to the public's basest fears of so-called frankenfoods, without trying to analyze (on its web site, for example) the scientifically based pros and cons of genetically modified foods. Rather, only potentially harmful consequences of genetic engineering are listed, without critical assessment. To further scare the public, frankenfoods are symbolized by a gigantic three-dimensional corncob, equipped with open, dark red lips and aggressively displayed fangs. Emotions and perhaps an undeclared political agenda seem to be the only factors driving this virulent opposition to biotechnology. I made it clear at the beginning of this chapter that I question the potential takeover of worldwide plant food production by a few multinational corporations. Yet I remain convinced that, outside of corporate profits, wisely applied plant biotechnology is the way of the future. In the end, science, but not violence or politics or profit, must be the ultimate arbiter of the validity of genetically modified plants in human nutrition and welfare.

Cesium Chloride Density Gradients and Their Use in DNA Analysis

℘

Cesium chloride (CsCl) is a very dense salt. It is also extremely soluble in water, and highly concentrated solutions (e.g., 6 mol/L) are used to run density gradients. The technique of CsCl equilibrium density gradient centrifugation for DNA analysis was invented in 1957 by Meselson and Stahl. The underlying idea is that Cs^+ ions (always neutralized by Cl^- ions) will sediment to the bottom of a tube when subjected to high g-forces. These forces can be achieved if the tube is spun in a centrifuge rotor at speeds of at least 30,000 rpm. The sedimentation of the Cs^+ ions is constantly counteracted by thermal diffusion, which forces these ions to move randomly in the tube. After enough time (typically 60 hours, or less, depending on the g-force applied, which depends on rotor speed and size), an equilibrium is established between sedimentation and thermal diffusion; a *gradient* thus forms in the tube, in which the concentration of Cs^+ is highest at the bottom of the tube and lowest at the top. This gradient is perfectly linear. DNA from all organisms has a density of about 1.7. Further, this density is directly dependent on the DNA (G + C) content, which has a characteristic value for each organism but differs among diverse kinds of plants animals, bacteria, and viruses. The higher the G + C content, the higher the density. Therefore, DNA molecules with different G + C contents will band (literally float) in the gradient at a place where they encounter their own density, at a certain CsCl concentration. They cannot sediment farther because a higher CsCl concentration (a higher density) will push them up; thermal diffusion, which would randomize the movement of these molecules, is exactly counteracted by g-forces, which push them down. This makes DNA bands quite narrow, and DNAs with different G + C contents can thus be separated. The CsCl gradient method is so sensitive that density values, up to at least the third decimal, can be determined with great accuracy. For example, the gradient profile in figure A1.1 shows the separation between *Arabidopsis* DNA (d = 1.696), *E. coli* DNA (d = 1.710), and *Micrococcus lysodeikticus* DNA (d = 1.731). Density (understood as the ratio between the density of water and that of another substance) is a dimensionless parameter, but these numbers are often expressed in terms of specific gravity, as in this graph. Specific gravity (ρ) is measured in gm/cm^3 and has the same absolute value as density. Single-stranded (denatured) DNA has a higher density than double-stranded DNA because the single strands "ball up" and thus occupy less space. At constant mass, single-stranded DNA will have a specific gravity (density) higher than that of double-stranded DNA, because ρ = mass/volume.

The width of DNA bands in CsCl gradients is proportional to the molecular mass (size) of the DNA; the higher the mass, the narrower the peak. The forces of diffusion scatter low-molecular-mass molecules farther than high-molecular-mass ones. For this reason, ultrasonication (shearing) of

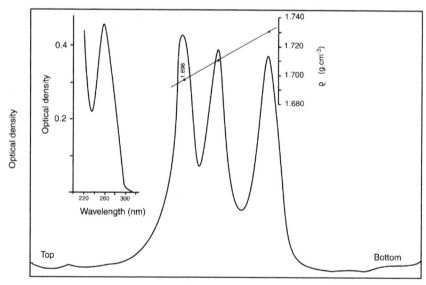

FIGURE A1.1.
CsCl banding and separation of DNA, from left to right, *Arabidopsis* (d = 1.696), *E. coli* (d = 1.710), and *M. lysodeikticus* (d = 1.731). *Inset*: UV spectrum of purified *Arabidopsis* DNA.

DNA molecules produces a broader peak in the CsCl gradient, but it should be centered on the same (average) density for molecules of homogeneous G + C content.

CsCl gradients are rarely used today for analytical purposes but are still used by some for the preparation of purified DNA. RNA sinks to the bottom and proteins float on top of a CsCl gradient that can band DNA.

Hybridization Experiments
with DNA

℮

The two strands of DNA are complementary, in the sense that an adenine (A) in one strand will always face and interact by hydrogen-bonding with a thymine (T) in the other strand. The same holds true for guanine (G) and cytosine (C). The strands of DNA can be separated by agents that destroy hydrogen bonds, such as high temperature (that of boiling water, 100°C, will denature DNA rapidly) or high pH (pH 11 is sufficient). Once the two strands of a DNA molecule are separated in solution, if the temperature (or pH) is brought down to a suitable point, the two strands can reassociate. They do this by "probing" one another for correct complementarity—that is, base-pairing. Perfect base-pairing between all the bases is ensured if reassociation is not too rapid and is allowed to occur several degrees below the temperature at which the double helix starts to dissociate. The rate at which the reassociation takes place can be measured. This is because single-stranded DNA absorbs more UV light at 260 nm than double-stranded DNA. What is measured is thus the drop in UV light absorbance as a function of time. If very small amounts of a test DNA are used, as in P_0t curves (see chapter 1), radioactive DNA must be used for detection. In that case, aliquots of reassociating DNA are harvested at given time intervals and run through a hydroxylapatite column. Hydroxylapatite has a greater affinity for double-stranded DNA than for its denatured, single-stranded version, and it is thus possible to elute one before the other. Thus the fraction of double-stranded (reassociated) DNA can be easily measured as a function of time.

Britten and Kohne (1968) demonstrated that DNA reassociation kinetics follows a second-order (sigmoidal) function. This was expected, because DNA reassociation kinetics involves two molecules of single-stranded DNA, hence the shape of the graphs. Figure A2.1 shows the results they obtained; these are C_0t curves obtained with DNAs from various sources. Low-complexity DNA molecules (such as phage DNA) reassociate very rapidly because the single strands do not have to "search" for one another very much. If complexity is high, as in nonrepetitive eukaryotic DNA, reassociation is orders of magnitude slower, because sequence heterogeneity is very high and most collisions between single strands will not result in a perfect double helix. Each single strand has to "wait" until it collides with its proper complement to form a perfect double-stranded molecule. Thus, the higher the complexity of DNA (the larger the number of genes it codes for), the higher the time it takes to reassociate.

This technique was then applied to look for homology between DNA molecules from different sources (P_0t curves). In the case of P_0t curves, the DNA to be analyzed for sequence complementarity (that is, homology with another DNA) is radiolabeled and mixed with a very large excess of the DNA it is suspected to be homologous with. This excess DNA is called the driver. If there is homology, the driver DNA will speed up the reassociation of the DNA to be analyzed because together they will form double-stranded molecules quickly. If there is no homology, the DNA to be analyzed will reassociate slowly because it remains at a low concentration as the driver cannot reassociate with it. This is clearly shown in chapter 1 (fig. 1.5), where the test DNA reassociated slowly in the presence of *Micrococcus lysodeikticus* DNA (no homology) but reassociated much more quickly in the presence of BC2Y DNA (definite homology).

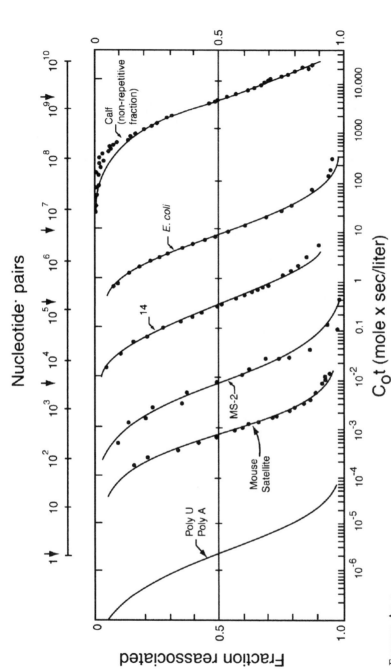

FIGURE A2.1.

$C_0 t$ curves. From left to right, reassociation kinetics of the synthetic polymers poly U and poly A, mouse satellite DNA, bacteriophage MS-2 RNA, bacteriophage T_4 DNA, *E. coli* DNA, and nonrepetitive calf DNA (From Britten, R. J., and D. E. Kohne. 1968. Repeated sequences in DNA. *Science* 161:529–540, figure 2. Copyright American Association for the Advancement of Science. Reprinted by permission of the publisher.)

APPENDIX 3

What Classical Genetics Says About Transformation Experiments

When diploid organisms acquire a new gene by transformation, often a single copy is picked up, but sometimes two or more. It is now known that the transforming genes are integrated randomly in the recipient chromosome, but it is not inconceivable (although extraordinarily unlikely, given the size of eukaryotic genomes) that a "correcting" gene, such as claimed in Ledoux's results, would simply replace the mutant locus or be very closely linked to it. Classical genetics can help unravel some different possibilities.

Following are four simplified examples that are applicable to *Arabidopsis* transformation claims, which we now know were artifactual. In these examples, I will use hypothetical chromosomes named X, Y, and Z (without reference to sex chromosomes) and I will call *a* a recessive lethal mutation and A^* a dominant allele contributed by foreign DNA taken up by the organism. The recipient lethal mutants are homozygous recessive for the *a* locus on chromosome Z and are thus Z^aZ^a. Chromosomes X and Y do not carry the *a* locus and could be any other chromosome, except Z.

Example 1: A Single Gene A^* Is Incorporated and Integrated Randomly in Chromosome X

In this case, the organism has the genotype $XX^{A^*}Z^aZ^a$, where X denotes the regular chromosome X, and X^{A^*} represents the other X, having acquired a copy of A^* by transformation. The gametes produced by this individual are thus of two types, XZ^a and $X^{A*}Z^a$. Upon selfing, this cross being the equivalent of a monohybrid cross, the phenotypic ratio will be 3 A* — aa to 1 aa or, in other words, 3/4 of the progeny will be corrected and 1/4 will consist of the original lethal mutants.

In a testcross with the homozygous recessive, that is, $XX^{A*}Z^aZ^a$ crossed with XXZ^aZ^a, one will expect a phenotypic ratio of 1 aa to 1 A*—, or 50 percent lethal mutants and 50 percent corrected. It is said that the *a* and *A** alleles *segregate*. These results were not obtained by Ledoux et al.

Example 2: Two Copies of Gene A* Are Integrated Randomly in the Recipient Genome

Here, I will assume that each copy of the A* gene is integrated within two different chromosomes, one in chromosome X and one in chromosome Y. If both copies are integrated side by side in the same chromosome, segregation patterns as in example 1 will be observed. If the two copies are on the same chromosome, but not right next to each other, segregation patterns will depend on the distance separating the two copies, but segregation of A* and a will still occur to some extent. This case will not be considered here.

Let us consider individual $XX^{A*}YY^{A*}Z^aZ^a$, which will produce four types of gametes because this is the equivalent of a dihybrid cross. The gametes are XYZ^a, $X^{A*}YZ^a$, $XY^{A*}Z^a$, and $X^{A*}Y^{A*}Z^a$. Upon selfing, it can be found by solving the cross that the phenotypic ratio is 15 A*—aa to 1 aa, the result expected in a case of gene duplication, which this is. Thus, the A* and *a* alleles segregate here, too. In a testcross with $XXYYZ^aZ^a$, the segregation will be 3 A*—aa to 1 aa. These results were not observed by Ledoux et al.

Example 3: The A* Gene Is Present in One Copy and Is Tightly Linked to or Replaces the *a* Locus

In this case, the corrected mutant can be written $Z^{A*/a} Z^a$ and will produce $Z^{A*/a}$ and Z^a gametes. This is a typical monohybrid cross, which upon selfing, will generate a 3 to 1 phenotypic ratio between the A*— and aa individuals. In a testcross, the ratio will be 1 A*a to 1 aa. Again, segregation is expected here, too, and again, this was not seen by Ledoux et al.

Example 4: The A* Gene Is Present in Two Copies That Are Tightly Linked to or Replace Both *a* Loci

Here, the individual will be homozygous $X^{A*/a}Z^{A*/a}$ and will of course produce 100 percent dominant offspring upon selfing; the *a* locus will no longer segregate and no mutants will appear in the progeny. This is what Ledoux et al. observed: no segregation. In other words, their results represent the most unlikely possibility, because this transformation model would assume that the two mutant alleles must be strongly linked to the two correcting genes.

In a testcross with $Z^{a}Z^{a}$, one would expect 100 percent $Z^{A*/a}Z^{a}$, which will also show a dominant phenotype. This was also observed by Ledoux et al., although lethal phenotypes appeared in more advanced generations (see chapter 2).

Plasmids, Transposons, and Horizontal Gene Transfer in Bacteria

ℰ

Plasmids are circular, double-stranded DNA molecules, independent of the chromosome, found in many bacterial strains. They have never been detected in eukaryotes, except in some fungi. Their sizes range from very small (about 2 kbp) to very large (up to 400 kbp), one-tenth the size of a "typical" bacterial chromosome. They code for functions that are not normally necessary for the survival of the bacteria harboring them. They code for accessory functions, useful or even essential in some cases, such as, for example, antibiotic resistance, heavy metal resistance, fixation of atmospheric nitrogen, and, of course, tumorigenesis in plants.

Many types of plasmids are promiscuous, meaning that they can be exchanged between closely and even distantly related bacterial species. These exchanges occur naturally in, for example, hospital settings (where human pathogens can acquire antibiotic resistance), sewers, and soil. Bacteria use three modes of DNA acquisition or exchange: conjugation, transformation, and transduction. Transduction consists of bacteriophage-mediated gene transfer but does not normally apply to plasmids. Plasmids can be acquired by transformation, a process of uptake of DNA from the environment by live bacteria after it has been released from dead, lysed bacteria. DNA can in some instances be released into the environment by living bacteria; for example, *Thermoactinomyces vulgaris* spontaneously releases DNA when its spores germinate.

Conjugation is a much more elaborate mechanism that bears a resemblance to sexual exchange in eukaryotes. Here, a "male" bacterial cell establishes a cytoplasmic bridge with a "female" via the formation of a conjugation tube. Plasmid DNA, in single-stranded form, can then travel through the

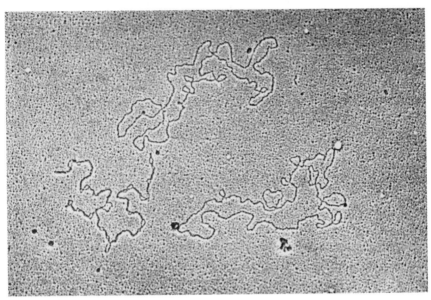

FIGURE A4.1.
Electron micrograph of plasmid R6K. One molecule is supercoiled and the two others are in the relaxed configuration.

conjugation tube, penetrate the female, and reform a double helix by replication. The male thus does not lose its plasmid, since the other strand is left behind and replicates as well. The same process also allows exchange of the chromosome. Bacteria do not have to be closely related to undergo plasmid conjugation, in contrast to eukaryotes, where interspecies exchange of genetic material is extremely rare. Indeed, the lack of cross-fertility is one of the defining characteristics of distinct species.

Transposons, also known as jumping genes, are pieces of DNA that code for their own mobility in the genome in which they reside. They contain a gene coding for a transposase, which is the enzyme that will cut them out of a particular location in a genome and cause their reinsertion elsewhere in that genome. Transposons insert randomly and often cause mutations when they transpose because they can interrupt genes in the process of either excision or insertion. They are ubiquitous, occurring in organisms from bacteria to humans. In prokaryotes, they often carry a gene coding for antibiotic resistance or other conditionally useful traits. Bacterial plasmids acquire multiple resistances to antibiotics simply by accumulating transposons. This mechanism is a curse for humans because it leads to the appearance of multidrug resistance in pathogenic bacteria. For geneticists, transposons are a tool for the same reason; they mutate genes randomly by interrupting them, can be selected for thanks to their antibiotic resistance, and can be located in the interrupted genes via restriction endonuclease and Southern blot analysis.

The T-DNA from the Ti plasmid is not mobilized to plant cells the way a transposon is (see chapter 3). The pTi is, however, a conjugative plasmid, and it can be mobilized from one *A. tumefaciens* cell to another in the presence of opines. The phenotypic effects of the *shi* and *roi* loci in the T-DNA, as well as some of the functions of the pTi *vir* genes (see chapter 3) were determined by transposon insertion mutagenesis.

Liposomes

Liposomes are microscopic lipid vesicles produced by dispersing phospholipids (in general) in an aqueous phase, in which they are insoluble. Phospholipids are hydrophilic (polar) on the charged (phosphate) ends and hydrophobic on the other (lipid) ends. Such molecules are called amphiphilic because of their affinity for both oil and water. These amphiphilic molecules dispersed into water can spontaneously form spherical bilayers, with their polar groups oriented toward the external and trapped (internal) aqueous phases. These bilayers imitate genuine cell membranes and can fuse with them. The trapped aqueous phase will contain the solutes initially present there before vesicle formation. If the liposomes are large enough, they can trap dissolved nucleic acids. There are basically two types of liposomes: multilamellar, in which several concentric bilayers are present within the liposome, and unilamellar, in which a single bilayer bounds the internal aqueous phase. It has been shown that unilamellar liposomes are better carriers of DNA and RNA for intracellular delivery. The surface charge of liposomes can be positive, neutral, or negative, depending on the electric charge of the lipids or other molecules used in their composition. It is generally agreed that negatively charged liposomes are more efficient for DNA and RNA delivery into plant protoplasts. Because the surface charge of protoplasts is also negative, calcium ions must be present in the incubation medium to block negative charges and enhance the effect of the fusogen (such as polyethylene glycol, PEG), which brings about the transfer of the DNA or RNA molecules from the liposomes to the protoplasts.

So-called cationic liposomes are provided preformed by manufacturers and simply need to be mixed with the nucleic acid in solution, with which they spontaneously form complexes of a nonvesicular nature. These complexes are of cylindrical geometry, with no bilayer formation. Each DNA molecule is thus coated with a cationic lipid monolayer, and these columns form hexagonal assemblies, with six complexes surrounding a central one.

FIGURE A 5.1.
Photomicrograph of multilamellar liposomes containing a fluorescent dye–DNA complex.

APPENDIX 6

Cell Electroporation

As its name indicates, the technique of cell electroporation consists in inducing pore formation in living cells through the action of an electric field. These pores are repaired quickly after the end of the electric discharge and, while open, make cells permeable to a whole host of compounds, including DNA.

Poration occurs only after the membrane breakdown voltage has been reached. The electric field strength necessary to reach this voltage can be calculated using the simplified Laplace equation:

$$E = V/1.5r,$$

where V (in volts) is the breakdown voltage (equal to approximately 1 V for living cells), r (in cm) is the radius of the cell, and E is the electric field strength in volts/cm. E thus depends on both the set voltage and the distance between the electrodes. Most plant protoplasts experience membrane breakdown around 250 V/cm.

Once the breakdown voltage has been reached or exceeded, the amount of compound (such as DNA or RNA) that will be internalized depends on the electrical energy (in joules, J) dissipated in the system per unit volume (EDV). If the pulse is delivered by a capacitance, the energy W is

$$W = \frac{1}{2}(CV^2),$$

where C is the value of the capacitance in farads and V is the set voltage in volts. In the case of a square wave pulse, the energy is

$$W = VIt = I^2Rt = (V^2/R)t,$$

where V is the set voltage in volts, I is the current in amperes, R is the resistance in ohms, and t is pulse time in seconds. To calculate EDV, W is divided by the volume of the protoplast suspension. Typical EDV values leading to electrotransformation are in the range of 100 J/ml.

Notes

Chapter 1

1. The correspondence between Lucien Ledoux and Mary-Dell Chilton was clearly not meant to remain private, as their letters were carbon-copied to several other investigators. I believe that further dissemination of these letters to others directly involved in the field was an excellent initiative.

2. The results published by the Tübingen group came from two doctoral dissertations, one by Waltraud Gradmann-Rebel and the other by Birgit Leber, both under the direction of Vera Hemleben. Their dissertations, of course, contained much more information than what was published. This additional result is reported in their work: Glycosylated T4 DNA produces the appearance of a high-density peak, whereas DNAs from *Streptomyces griseus, E. coli, B. subtilis,* and *Pneumococcus* do not. These bacterial DNAs, when fed to *Matthiola*, are degraded and reutilized for endogenous DNA synthesis and produce no intermediate-density DNA peak.

Chapter 2

1. This fact was later acknowledged by Hess himself. He observed that on one given plant, the phenotype could change from white flowers to heavy anthocyanin accumulation in the corolla and then reverse, depending on temperature and light intensity. These changes were also seasonal. In addition, he noted that viral infections (which were not considered earlier) also influenced anthocyanin production in his plants (Hess 1980).

2. One may wonder why it was so important to use plant mutants unable to perform a certain function (such as pigment or thiamin synthesis). Why not go directly for an *added* function, such as disease or herbicide resistance? The answer is twofold: First, disease or herbicide resistance genes were either not known or not well characterized at the time; second, these experiments were aimed at determining the *feasibility* of actually having exogenous DNA expressed in recipient plant cells. For this, a reasonable approach was to *correct* a known mutation with exogenous DNA known to contain the gene(s) performing that very function.

3. This procedure is fundamental to the science of genetics and is called *selection*. Selection results when cells or organisms are placed under conditions that do not allow them to survive (for example, in the presence of an antibiotic, in the case of bacteria). If any of them do survive, there is an excellent chance that they have acquired a new trait through mutation, a heritable change in the DNA. In the *Arabidopsis* experiments, the selective agent is thiamin or one of its precursors; in its absence, mutants will die. If incorporated foreign DNA has brought with it the gene(s) necessary for thiamin synthesis, these transformed mutants will survive.

4. There is a very sad aspect to that story. Other than the fact that, one more time, a wild claim was proven wrong, this potential treatment for galactosemia in children had drawn the attention of many medical geneticists throughout the world. My former genetics professor, René Thomas, the lambdologist, and a colleague, a physician specialized in genetic diseases, were among them. They, like others, tried to reproduce Merril's data and failed. Many people were deeply disappointed.

5. Hess had not given up on his studies on flower color in *Petunia*. He applied his "pollen soaking in DNA" technique to the old flower color transformation problem that he had tackled 10 years earlier. In these new studies, several unexpected phenotypes showed up again. These were categorized in four different classes, and two of these classes were assigned the designation "heterozygote" based on anthocyanin distribution and intensity. These "heterozygotes" were selfed and testcrossed but failed to give Mendelian ratios. For example, in F3, no homozygous plants (with totally red flowers) were recovered. This was blamed on the lethality of the "red" allele in these plants. One wonders how it would then be possible to obtain these plants at all. Yet they do exist, because the donor DNA was extracted from them. Similarly, testcrosses, which should have generated a 1:1 ratio between heterozygous plants and those bearing white flowers, gave three types of plants, not two. The same difficulties appeared as before; without clear-cut phenotypic categories, it is impossible to draw clear-cut conclusions as to whether these traits undergo Mendelian inheritance.

Chapter 3

1. Some claims were right and temporarily rejected anyway. For example, Robb Schilperoort and his team in Leiden had a difficult time convincing the scientific community that detection of opines in plant tissues was a sign of crown gall transformation. Several reports, the earliest one of which is cited here (Wendt-Gallitelli and Dobrigkeit 1973), contested the fact that opines were a definitive marker for crown gall transformation and criticized Robb's work (his Ph.D. thesis, in fact) and the work of others. Schilperoort's approach was eventually vindicated, and his own efforts resulted in a universally accepted, rapid enzymatic assay for crown gall transformation diagnosis (Otten and Schilperoort 1978). Why did other people initially disagree? The reasons range from analysis of tissues that were not crown gall (they were habituated tissues—that is, cells that had spontaneously become hormone

independent and that, of course, would not contain opines) to color-blindness of the investigator (opine spots in electrophoregrams are yellow but turn blue with time). Also, supposedly habituated tobacco tissue isolated in France in 1946, producing octopine and displaying phytohormone independence, was widely circulated and puzzled researchers for years. It was finally demonstrated that this tissue was in fact genuine crown gall (Yang et al. 1980). An apparent mislabeling thus complicated the interpretation of crown gall transformation for a long time.

2. Plant scientists can have a good sense of humor, sometimes bittersweet for others. When Jef Schell was still heading two research groups at once (one in Ghent, Belgium, and the other one in Cologne, Germany), it was not unusual for his coworkers, speaking at meetings, to show an introductory slide portraying two soccer teams (22 people), each composed of a portion of Schell's binational laboratories. The public could then easily understand the magnitude of their research efforts. Also, Jef Schell was a fair laboratory chief: He played the first half-time with one team and the second with the other team.

Chapter 4

1. Some will object to the idea that protoplasts are easy to produce and manipulate. However, well-established, reliable techniques have been in existence for a long time for the preparation and cultivation of, among others, tobacco protoplasts. Further, leaf mesophyll cells, often used for the preparation of protoplasts, are arrested in G0 and do not synthesize DNA. Hence, freshly isolated mesophyll protoplasts do not synthesize DNA either and do not reutilize the breakdown products of foreign DNA presented to them. This greatly simplified the interpretation of uptake experiments involving radiolabeled donor DNA.

Glossary

Agrobacterium: A genus of soil bacteria, including *A. tumefaciens*, *A. rhizogenes*, *A. rubi*, and *A. radiobacter*. *A. tumefaciens* is a plant pathogen that causes crown gall disease, which is characterized by tumor formation at the site of infection.

Allele: One of two or more alternative forms of a single gene.

Auxotrophic mutation: A mutation affecting an organism's ability to make a nutrient essential for its survival.

Axenic: An axenic organism is one that lives or is cultivated, usually under laboratory conditions, in the complete absence of other organisms. Axenic plants, for example, do not harbor bacterial, fungal, viral, or other living contaminants.

Bacteriophage (phage): A virus that infects bacteria.

Callus: A mass of undifferentiated plant cells cultivated on synthetic medium in the laboratory.

Catabolism: The breaking down, or degrading, of compounds by a cell or organism so that they can be further used in metabolism.

Chimeric: Composed of parts with different origins. A chimeric gene contains a promoter from one origin (e.g., plant) and a coding sequence from another (e.g., bacterial).

Chromatography: A technique used to separate compounds that bind differently to a matrix as they flow in a liquid or a gas. Single-stranded DNA can be separated from double-stranded DNA by liquid chromatography.

Cloning: The generating of many identical copies of a DNA segment by replication in a living host such as *E. coli*.

Codon: A set of three nucleotide bases in mRNA specifying a particular amino acid. A nonsense codon specifies a translational stop—that is, termination of protein synthesis.

Complementation: In the context here, the correction of a mutation by the addition of a wild-type, functional gene (allele).

Concatenation: Several independent DNA molecules forming a multimeric complex by covalent attachment.

Conjugation: Bacterial sex; the exchange of DNA between bacteria (see appendix 4).

Constitutive gene expression: Qualifies a gene that is transcribed at all times, or that is always "on."

Cosuppression: Also called "gene silencing," the inhibition of transgene expression when present at high multiplicity.

CsCl density centrifugation: A centrifugation technique using cesium chloride that allows the separation of DNA molecules according to their G + C content (see appendix 1).

Direct gene transfer: The action of introducing DNA molecules into organisms using physical or chemical helpers.

Electroporation: The formation of pores in a cell membrane through the action of an electrical discharge (see appendix 6).

Endocytosis: The process by which a cell folds its membrane, invaginates it, and finally pinches off part of it so that it becomes an internal vesicle. Many types of animal cells (but not plant cells) use this process to ingest solutes present in their environment.

Endogenous: Originating within an organism. Endogenous DNA is that which is naturally present in a cell or organism.

Eukaryotes: A class of cells or organisms whose cells possess a nucleus. Plants and animals are eukaryotes.

Exogenous: Coming from outside an organism. Exogenous DNA is DNA supplied externally to a cell or organism. In this context, the word is synonymous with foreign, heterologous, and donor.

Fusogen: A chemical compound that has the ability to fuse protoplasts.

Gene silencing: "Turning off," or inactivation, of genes known to be present in an organism. Gene silencing may have multiple causes.

Genome: The set of all genes present in a cell or organism.

Gram-positive: Describes bacteria that stain purple (or positive) with the Gram stain technique.

Homologous: In the case of DNA, describes molecules that share the same base sequence.

Homologous recombination: The process by which two DNA molecules that share homologous sequences break and join.

Horizontal (or lateral) gene transfer: Transfer of genetic material between unrelated organisms. Such transfer rarely occurs in nature, except via viruses or via plasmids in bacteria. In contrast, vertical gene transfer takes place between progenitor(s) and daughter cells or organisms. Sexual reproduction is an example of vertical gene transfer.

Hybridization: A technique used to determine whether DNA (or RNA) molecules from different sources share sequence homology (see appendix 2).

Induction of gene expression: The action of "turning genes on." Many genes are normally "off" (repressed) and can be induced through a regulatory cascade often starting with an external stimulus.

kbp: Kilo (or thousand) base pairs.

Linkage: Describing genes located on the same chromosome.

Liposomes: Lipid-containing vesicles that can be artificially created (see appendix 5).

Mbp: Million base pairs.

Meiotic drive: The selection during meiosis of certain gametes possessing a particular combination of alleles.

Microarrays: Arrays of up to thousands of DNA sequences (arranged on a microchip), which can be hybridized with cDNAs (see reverse transcriptase) isolated from cells. This technique allows the *en masse* study of gene expression and regulation in organisms.

Nucleosome: A eukaryotic feature of DNA arrangement. The DNA double helix becomes folded in a supercoiled fashion around an octamer of histone proteins in the nucleus.

Oncogene: A cancer-causing gene.

Operon: A set of clustered genes, all involved in a given metabolic pathway, under the control of a single promoter.

Phytohormones: Plant hormones necessary for growth and development. The only phytohormones considered here are auxins (favoring root formation) and cytokinins (favoring shoot formation).

Plasmid: A piece of covalently closed, circular DNA existing and replicating independently from the chromosome (see appendix 4).

Polyadenylation: Attachment of a sequence of multiple adenines at the 3′ end of a eukaryotic mRNA.

Polycistronic mRNA: A single mRNA molecule representing several genes, usually the transcript of an operon.

Protoplast: A plant, bacterial, or fungal cell devoid of cell wall as a result of either mutation or chemical digestion.

Prokaryotes: Cells devoid of a nucleus. Examples are bacteria and archaebacteria.

Pseudogene: A DNA sequence that looks like a gene but is not expressed for any of a variety of reasons, such as absence of promoter.

Restriction endonuclease: An enzyme of bacterial origin that cleaves DNA molecules at a specific base sequence.

Reverse transcriptase: An enzyme able to synthesize a DNA copy from an RNA template. These DNA copies are called cDNAs.

Segregation: In diploid or partially diploid organisms, separation of alleles carried by different chromosomes or DNA molecules into different gametes or progeny cells.

Selection: The action of placing living cells or organisms under conditions that favor the multiplication or absence of a particular genotype.

Selfing: In plants, the act of self-fertilizing and producing progeny.

Somatic mutation: A mutation affecting a somatic cell only and therefore not transmissible to the progeny.

Sonication: The action of submitting an object or solution to ultrasound. Sonication of DNA in solution results in breakage of the molecules. Also called ultrasonication.

Specific radioactivity: Amount of radioactivity per unit mass.

Strain: A line of cells, all presumably descended from a single ancestor, all possessing the same set of physical, physiological, and genetic characteristics. A strain is also a clone if all members have descended from a single ancestor.

Supercoiling: The ability of DNA molecules to adopt an order of coiling higher than the double helix by twisting either around themselves or around a protein core. A photographic example is given in appendix 4.

Testcross: Sexually crossing an individual of known dominant phenotype (but of unknown genotype, which may be homozygous or heterozygous for the dominant marker gene) to a homozygous recessive individual.

Thiamin: Vitamin B_1.

Transformation: In the context here, introduction of new, foreign genes into an organism by transfer of DNA.

Transgene: A foreign gene introduced into a cell or organism by transformation.

Transgenesis: The technology allowing transformation by transfer of DNA.

Transposon: Also called a jumping gene. A segment of DNA able to excise itself from a chromosomal locus and reintegrate itself elsewhere in the genome. Transposons exist in both prokaryotes and eukaryotes (see appendix 4).

References

Akella, V., and P. F. Lurquin. 1993. Expression in cowpea seedlings of chimeric transgenes after electroporation into seed-derived embryos. *Plant Cell Rep.* 12:110–117.

Albright, L. M., M. F. Yanofsky, B. Leroux, M. A. Dequin, and E. W. Nester. 1987. Processing of the T-DNA of *Agrobacterium tumefaciens* generates border nicks and linear, single-stranded T-DNA. *J. Bacteriol.* 169:1046–1055.

An, G. 1985. High efficiency transformation of cultured tobacco cells. *Plant Physiol.* 79:568–570.

An, G., B. D. Watson, and C. C. Chiang. 1986. Transformation of tobacco, tomato, potato and *Arabidopsis thaliana* using a binary Ti vector system. *Plant Physiol.* 81:301–305.

An, G., B. D. Watson, S. Stachel, M. P. Gordon, and E. W. Nester. 1985. New cloning vehicles for transformation of higher plants. *EMBO J.* 4:277–284.

Anandalakshmi, R., G. J. Pruss, X. Ge, R. Marathe, A. C. Mallory, T. H. Smith, and V. B. Vance. 1998. A viral suppressor of gene silencing in plants. *Proc. Natl. Acad. Sci. USA* 95:13079–13084.

Anker, P., and M. Stroun. 1968. Bacterial nature of radioactive DNA found in tomato plants incubated in the presence of bacterial DNA-^{3}H. *Nature* 219:932–933.

Anonymous. 1974. Transferring foreign genes to plants. *Nature* 249:9.

Antonelli, N. M., and J. Stadler. 1990. Genomic DNA can be used with cationic methods for highly efficient transformation of maize protoplasts. *Theor. Appl. Genet.* 80:395–401.

Aoki, S., and I. Takebe. 1969. Infection of tobacco mesophyll protoplasts by tobacco mosaic virus ribonucleic acid. *Virology* 39:439–448.

———. 1975. Replication of tobacco mosaic virus RNA in tobacco mesophyll protoplasts inoculated in vitro. *Virology* 65:343–354.

Barton, K. A., A. N. Binns, A. J. M. Matzke, and M.-D. Chilton. 1983. Regeneration of intact tobacco plants containing full length copies of genetically engineered T-DNA, and transmission of T-DNA to R1 progeny. *Cell* 32:1033–1043.

Beggs, J. D. 1978. Transformation of yeast by a replicating hybrid plasmid. *Nature* 275:104–109.

Bendich, A. 1972. Effect of contaminating bacteria on the radiolabeling of nucleic acids from seedlings: False DNA "satellites." *Biochim. Biophys. Acta* 272:494–503.

Bendich, A., and P. Filner. 1971. Uptake of exogenous DNA by pea seedlings and tobacco cells. *Mutation Research* 13:199–214.

Bevan, M. 1984. Binary *Agrobacterium* vectors for plant transformation. *Nucleic Acids Res.* 12:8711–8721.

Bevan, M. W., R. B. Flavell, and M.-D. Chilton. 1983. A chimaeric antibiotic resistance gene as a selectable marker for plant cell transformation. *Nature* 304:184–187.

Bianchi, F., and H. G. Walet-Foederer. 1974. An investigation into the anatomy of the shoot apex of *Petunia hybrida* in connection with the results of transformation experiments. *Acta Bot. Neerl.* 23:1–6.

Bodley, J. H. 1996. Cultural Anthropology: Tribes, States and the Global System, 2nd ed. Mountain View, CA: Mayfield.

Bomhoff, G., P. M. Klapwijk, H. C. M. Kester, R. A. Schilperoort, J. P. Hernalsteens, and J. Schell. 1976. Octopine and nopaline synthesis and breakdown genetically controlled by a plasmid of *Agrobacterium tumefaciens*. *Mol. Gen. Genet.* 145:177–181.

Britten, R. J., and D. E. Kohne. 1968. Repeated sequences in DNA. *Science* 161:529–540.

Caboche, M., and P. F. Lurquin. 1987. Liposomes as carriers for the transfer and expression of nucleic acids into higher plant protoplasts. In L. Packer and R. Douce, eds. *Methods in Enzymology*, vol. 148, pp. 39–45. San Diego: Academic Press.

Caplan, A., L. Herrera-Estrella, D. Inzé, E. Van Haute, M. Van Montagu, J. Schell, and P. Zambryski. 1983. Introduction of genetic material into plant cells. *Science* 222:815–821.

Cardi, T., V. Iannamico, F. D'Ambrosio, E. Fillipone, and P. F. Lurquin. 1992. *Agrobacterium*-mediated genetic transformation of *Solanum commersonii* Dun. *Plant Science* 87:179–189.

Chen, G.Y., A. J. Conner, J. Wang, A. G. Fautrier, and R. J. Field. 1999. Energy dissipation as a key factor for electroporation of protoplasts. *Mol. Biotech.* 10:209–216.

Chilton, M.-D., M. H. Drummond, D. J. Merlo, D. Sciaky, A L. Montoya, M. P. Gordon, and E. W. Nester. 1977. Stable incorporation of plasmid DNA into higher plant cells: The molecular basis of crown gall tumorigenesis. *Cell* 11:263–271.

Chilton, M.-D., R. K. Saiki, N. Yadav, M. P. Gordon, and F. Quetier. 1980. T-DNA from *Agrobacterium* Ti plasmid is in the nuclear DNA fraction of crown gall tumor cells. *Proc. Natl. Acad. Sci. USA* 77:4060–4064.

Chowrira, G. M., V. Akella, E. P. Fuerst, and P. F. Lurquin. 1996. Transgenic grain legumes obtained by in planta electroporation-mediated gene transfer. *Mol. Biotech.* 5:85–96.

Chowrira, G. M., T. D. Cavileer, S. K. Gupta, P. F. Lurquin, and P. H. Berger. 1998. Coat protein-mediated resistance to pea enation mosaic virus in transgenic *Pisum sativum* L. *Transgenic Research* 7:265–271.

Christen, A. A., and P. F. Lurquin. 1983. Infection of cowpea mesophyll protoplasts with cowpea chlorotic mottle virus (CCMV) encapsulated in large liposomes. *Plant Cell Rep.* 2:43–46.

Cocking, E. C. 1977. Genetic modification of plant cells: A reappraisal. *Nature* 266:13–14.

Cornelissen, M., and M. Vandewiele. 1989. Both RNA level and translation efficiency are reduced by anti-sense RNA in transgenic tobacco. *Nucleic Acids Res.* 17:833–843.

Crossway, A., J. V. Oakes, J. M. Irvine, B. Ward, V. C. Knauf, and C. K. Shewmaker. 1986. Integration of foreign DNA following microinjection of tobacco mesophyll protoplasts. *Mol. Gen. Genet.* 202:179–185.

Davey, M. R., E. C. Cocking, J. Freeman, J. Draper, N. Pearce, I. Tudor, J. P. Hernalsteens, M. De Beuckeleer, M. Van Montagu, and J. Schell. 1980. The use of plant protoplasts for transformation by *Agrobacterium* and isolated plasmids. In L. Ferenczy and G.L. Farkas, eds. *Advances in Protoplast Research*, pp. 425–430. New York: Pergamon Press.

De Block, M., J. Botterman, M. Vandewiele, J. Dockx, C. Thoen, V. Gosselé, R. N. Movva, C. Thompson, M. Van Montagu, and J. Leemans. 1987. Engineering herbicide resistance in plants by expression of a detoxifying enzyme. *EMBO J.* 6:2513–2518.

De Block, M., J. Schell, and M. Van Montagu. 1985. Chloroplast transformation by *Agrobacterium tumefaciens*. *EMBO J.* 4:1367–1372.

DeFramond, A. J., K. A. Barton, and M.-D. Chilton. 1983. Mini-Ti: A new vector strategy for plant genetic engineering. *Bio/Technology* 1:262–269.

Dekeyzer, R. A., B. Claes, R. M. U. De Rycke, M. E. Habets, M. Van Montagu, and A. B. Caplan. 1990. Transient gene expression in intact and organized rice tissues. *Plant Cell* 2:591–602.

DellaPenna, D. 1999. Nutritional genomics: Manipulating plant micronutrients to improve human health. *Science* 285:375–379.

Dellaporta, S. L. 1981. Studies on the interaction of plant protoplasts and liposomes: The development of a transformation system for higher plants. Ph.D. Dissertation, Worcester Polytechnic Institute.

Dellaporta, S. L., and K. L. Giles. 1979. A possible method for the high frequency transformation of plant protoplasts. *Abstracts of the Fifth International Protoplast Symposium*, p. 138. Szeged, Hungary.

Denhardt, D. T. 1966. A membrane-filter technique for the detection of complementary DNA. *Biochem. Biophys. Res. Comm.* 23:641–646.

Depicker, A. G., A. M. Jacobs, and M. C. Van Montagu. 1988. A negative selection scheme for tobacco protoplast-derived cells expressing the T-DNA gene 2. *Plant Cell Rep.* 7:63–66.

Depicker, A., and M. Van Montagu. 1997. Post-transcriptional gene silencing in plants. *Curr. Opin. Cell Biol.* 9:373–382.

Deroles, S. C., and R. C. Gardner. 1988a. Expression and inheritance of kanamycin resistance in a large number of transgenic petunias generated by *Agrobacterium*-mediated transformation. *Plant Mol. Biol.* 11:355–364.

———. 1988b. Analysis of the T-DNA structure in a large number of transgenic petunias generated by *Agrobacterium*-mediated transformation. *Plant Mol. Biol.* 11:365–377.

Deshayes, A., L. Herrera-Estrella, and M. Caboche. 1985. Liposome-mediated transformation of tobacco mesophyll protoplasts by an *Escherichia coli* plasmid. *EMBO J.* 4:2731–2737.

Dev, S. B., and Y. Hayakawa. 1999. Electroporation-mediated molecular transfer in intact plants. U.S. Patent #5,859,327.

Dieryck, W., J. Pagnier, M. C. Marden, V. Gruber, P. Bournat, S. Baudino, and B. Mérot. 1997. Human haemoglobin from transgenic tobacco. *Nature* 386:29–30.

Dillen, W., J. Engler, M. Van Montagu, and G. Angenon. 1995. Electroporation-mediated DNA delivery to seedling tissues of *Phaseolus vulgaris* L. (common bean). *Plant Cell Rep.* 15:119–124.

Dimitriadis, G. I. 1978. Translation of rabbit globin mRNA introduced by liposomes into mouse lymphocytes. *Nature* 274:223–224.

Doolittle, W.F. 1999. Phylogenetic classification and the universal tree. *Science* 284:2124–2128.

Doy, C. H., P. M. Gresshoff, and B. G. Rolfe. 1973a. Biological and molecular evidence for the transgenosis of genes from bacteria to plant cells. *Proc. Nat. Acad. Sci. USA* 70:723–726.

———. 1973b. Time-course of phenotypic expression of *Escherichia coli* gene Z following transgenosis in haploid *Lycopersicon esculentum* cells. *Nature New Biology* 244:90–91.

Drummond, M. H., M. P. Gordon, E. W. Nester, and M.-D. Chilton. 1977. Foreign DNA of bacterial plasmid origin is transcribed in crown gall tumours. *Nature* 269:535–536.

Feenstra, W. J., D. L. De Heer, and F. J. Oostindier-Braaksma. 1973. Negative results of treatment with bacterial DNA on "repair" of mutants of *Arabidopsis*. *Arabidopsis Inf. Serv.* 10:33.

Fraley, R. T., S. L. Dellaporta, and D. Papahadjopoulos. 1982. Liposome-mediated delivery of tobacco mosaic virus RNA into tobacco protoplasts: A sensitive assay for monitoring liposome-protoplasts interactions. *Proc. Natl. Acad. Sci. USA* 79:1859–1863.

Fraley, R. T., S. G. Rogers, R. B. Horsch, P. R. Sanders, J. S. Flick, S. P. Adams, M. L. Bittner, L. A. Brand, C. L. Fink, J. S. Fry, G. R. Galluppi, S. B. Goldberg, N. L. Hoffmann, and S. C. Woo. 1983. Expression of bacterial genes in plant cells. *Proc. Natl. Acad. Sci. USA* 80:4803–4807.

Fromm, M., L. P. Taylor, and V. Walbot. 1985. Expression of genes transferred into monocot and dicot plant cells by electroporation. *Proc. Natl. Acad. Sci. USA* 82:5824–5828.

Fukunaga, Y., T. Nagata, and I. Takebe. 1981. Liposome-mediated infection of plant protoplasts with tobacco mosaic virus RNA. *Virology* 113:752–760.

Garfinkel, D. G., R. B. Simpson, L. W. Ream, F. F. White, M. P. Gordon, and E. W. Nester. 1981. Genetic analysis of crown gall: Fine structure map of the T-DNA by site-directed mutagenesis. *Cell* 27:143–153.

Gaskell, G., M. W. Bauer, J. Durant, and N. C. Allum. 1999. Worlds apart? The reception of genetically modified foods in Europe and the U.S. *Science* 285:384–387.

Gheysen, G., P. Dhaese, M. Van Montagu, and J. Schell. 1985. DNA flux across genetic barriers: The crown gall phenomenon. In B. Hohn and E. S. Dennis, eds. *Genetic Flux in Plants*, pp. 11–47. Wien: Springer-Verlag.

Gielen, J., M. De Beuckeleer, J. Seurinck, F. Deboeck, H. De Greve, M. Lemmers, M. Van Montagu, and J. Schell. 1984. The complete nucleotide sequence of the TL-DNA of the *Agrobacterium tumefaciens* plasmid pTIAch5. *EMBO J.* 4:835–846.

Golds, T., P. Maliga, and H.-U. Koop. 1993. Stable plastid transformation in PEG-treated protoplasts of *Nicotiana tabacum. Bio/Technology* 11:95–97.

Gordon, J. W., G. A. Scangos, D. J. Plotkin, J. A. Barbosa, and F. H. Ruddle. 1980. Genetic transformation of mouse embryos by microinjection of purified DNA. *Proc. Natl. Acad. Sci. USA* 77:7380–7384.

Gradmann-Rebel, W., and V. Hemleben. 1976. Incorporation of T4 phage DNA into a specific DNA fraction from the higher plant *Matthiola incana. Z. Naturforsch.* 31c:558–564.

Graham, F. L., and A. J. van der Eb. 1973. A new technique for the assay of infectivity of human adenovirus 5 DNA. *Virology* 52:456–467.

Grandbastien, M-A., A. Spielmann, and M. Caboche. 1989. Tnt1, a mobile, retroviral-like transposable element of tobacco isolated by plant cell genetics. *Nature* 337:376–380.

Graves, A. C. F., and S. L. Goldman. 1986. The transformation of *Zea mays* seedlings with *Agrobacterium tumefaciens*: detection of T-DNA specific enzyme activities. *Plant Mol. Biol.* 7: 43–50.

———. 1987. *Agrobacterium tumefaciens*-mediated transformation of the monocot genus *Gladiolus*: Detection of expression of T-DNA-encoded genes. *J. Bacteriol.* 169:1745–1746.

Gronenborn, B., Gardner, R.C., Schaefer, S., and Shepherd, R.J. 1981. Propagation of foreign DNA in plants using cauliflower mosaic virus as vector. *Nature* 294:773–776.

Hain, R., H.-H. Steinbis, and J. Schell. 1984. Fusion of *Agrobacterium* and *E. coli* spheroplasts with *Nicotiana tabacum* protoplasts—Direct gene transfer from microorganism to higher plant. *Plant Cell Rep.* 3:60–64.

Hamilton, A. J., and D. C. Baulcombe. 1999. A species of small antisense RNA in postrancriptional gene silencing in plants. *Science* 286: 950–952.

Hansen, G., and M.-D. Chilton. 1999. Lessons in gene transfer to plants by a gifted microbe. *Curr. Top. Microbiol. Immunol.* 240:21–57.

Hanson, R. S., and M.-D. Chilton. 1975. On the question of integration of *Agrobacterium tumefaciens* deoxyribonucleic acid by tomato plants. *J. Bacteriol.* 124:1220–1226.

Hasezawa, S., T. Nagata, and K. Syono. 1981. Transformation of *Vinca* protoplasts mediated by *Agrobacterium* spheroplasts. *Mol. Gen. Genet.* 182:206–210.

Hemleben, V., N. Ermisch, D. Kimmich, B. Leber, and G. Peter. 1975. Studies on the fate of homologous DNA applied to seedlings of *Matthiola incana*. *Eur. J. Biochem.* 56:403–411.

Hernalsteens, J. P., L. Thia-Toong, J. Schell, and M. Van Montagu. 1984. An *Agrobacterium* transformed cell culture from the monocot *Asparagus officinalis*. *EMBO J.* 3:3039–3042.

Herrera-Estrella, A., M. Van Montagu, and K. Wang. 1990. A bacterial peptide acting as a plant nuclear targeting signal: The amino-terminal portion of *Agrobacterium* VirD2 protein directs a β-galactosidase fusion protein into tobacco nuclei. *Proc. Natl. Acad. Sci. USA* 87:9534–9537.

Herrera-Estrella, L., A. Depicker, M. Van Montagu, and J. Schell. 1983. Expression of chimaeric genes transferred into plant cells using a Ti-plasmid-derived vector. *Nature* 303:209–213.

Hess, D. 1972. Versuche zur Transformation an höheren Pflanzen: Nachweis von Heterozygoten in Versuchen zur Transplantation von Genen für Anthocyan-synthese bei *Petunia hybryda*. *Z. Pflanzenphysiol.* 66:155–166.

———. 1978. Genetic effects in *Petunia hybrida* induced by pollination with pollen treated with *lac* transducing phages. *Z. Pflanzenphysiol.* 119:132.

———. 1979. Genetic effects in *Petunia hybrida* induced by pollination with pollen treated with *gal* transducing phages. *Z. Pflanzenphysiol.* 93:429–436.

———. 1980. Investigations on the intra- and interspecific transfer of anthocyanin genes using pollen as vectors. *Z. Pflanzenphysiol.* 98:321–337.

Hess, D., and K. Dressler. 1984. Bacterial transferase activity expressed in *Petunia* progeny. *J. Plant Physiol.* 116:261–272.

Heyn, R. F., and R. A. Schilperoort. 1973. The use of protoplasts to follow the fate of *Agrobacterium tumefaciens* DNA on incubation with tobacco cells. *Colloques Internationaux C. N. R. S.* 212:385–395.

Hiei, Y., S. Ohta, T. Komari, and T. Kumashiro. 1994. Efficient transformation of rice (*Oryza sativa* L.) mediated by *Agrobacterium* and sequence analysis of the boundaries of the T-DNA. *The Plant J.* 6:271–282.

Hinchee, M. A. W., D. V. Connor-Ward, C. A. Newell, R. E. McDonnell, S. J. Sato, C. S. Gasser, D. A. Fischhoff, D. B. Re, R. T. Fraley, and R. Horsch. 1988. Production of transgenic soybean plants using *Agrobacterium*-mediated DNA transfer. *Bio/Technology* 6:915–922.

Hinnen, A., J. B. Hicks, and G. R. Fink. 1978. Transformation of yeast. *Proc. Natl. Acad. Sci. USA* 75:1929–1933.

Hoekema, A., P. R. Hirsch, P. J. J. Hooykaas, P. J. Schilperoort, and R. A. Schilperoort. 1983. A binary plant vector strategy based on separation of vir- and T-region of the *Agrobacterium tumefaciens* Ti-plasmid. *Nature* 303:179–180.

Hoffman, R. M., L. B. Margolis, and L. B. Bergelson. 1978. Binding and entrapment of high molecular weight DNA by lecithin liposomes. *FEBS Lett.* 93:365–368.

Hooykaas, P. J. J., P. M. Klapwijk, M. P. Nuti, R. A. Schilperoort, and A. Rörsch. 1977. Transfer of the *Agrobacterium tumefaciens* TI plasmid to avirulent agrobacteria and to *Rhizobium* ex planta. *J. Gen. Microbiol.* 98:477–484.

Horsch, R. B., R. T. Fraley, S. G. Rogers, A. Lloyd, and N. L. Hoffmann. 1984. Inheritance of foreign genes in plants. *Science* 223:496–498.

Horsch, R. B., J. E. Fry, N. L. Hoffmann, D. Eichholtz, S. G. Rogers, and R. T. Fraley. 1985. A simple and general method for transferring genes into plants. *Science* 227:1229–1231.

Hotta, Y., and H. Stern. 1971. Uptake and distribution of heterologous DNA in living cells. In L. Ledoux, ed. *Informative Molecules in Biological Systems*, pp. 176–186. New York: American Elsevier.

Huang, F., L. L. Buschman, R. A. Higgins, and W. H. McGaughhey. 1999. Inheritance of resistance to *Bacillus thuringiensis* toxin (Dipel ES) in the European corn borer. *Science* 284:965–967.

Hughes, B. G., F. G. White, and M. A. Smith. 1977. Fate of bacterial plasmid DNA during uptake by barley protoplasts. *FEBS Lett.* 79:80–84.

———. 1979. Fate of bacterial plasmid DNA during uptake by barley and tobacco protoplasts: II. Protection by poly-L-ornithine. *Plant Science Lett.* 14:303–310.

Iyer, V. N., H. J. Klee, and E. W. Nester. 1982. Units of genetic expression in the virulence region of a plant tumor-inducing plasmid of *Agrobacterium tumefaciens*. *Mol. Gen. Genet.* 188:418–414.

Jefferson, R. A., T. A. Kavanagh, and M. W. Bevan. 1987. GUS fusions: β-glucuronidase as a sensitive and versatile gene fusion marker in higher plants. *EMBO J.* 6:3901–3907.

Johnson, C. B., D. Grierson, and H. Smith. 1975. Expression of λ*plac5* DNA in cultured cells of a higher plant. *Nature New Biology* 244:105–107.

Jorgensen, R. A., R. G. Atkinson, R. L. S. Forster, and W. J. Lucas. 1998. An RNA-based information superhighway in plants. *Science* 279:1486–1487.

Kado, C. I., and P. F. Lurquin. 1976. Studies on *Agrobacterium tumefaciens*. V. Fate of exogenously added bacterial DNA in *Nicotiana tabacum*. *Physiol. Plant Pathol.* 8:73–82.

Klein, T. M., M. Fromm, A. Weissinger, D. Tomes, S. Schaaf, M. Sletten, and J. C. Sanford. 1988. Transfer of foreign genes into intact maize cells with high-velocity microprojectiles. *Proc. Natl. Acad. Sci. USA* 85:4305–4309.

Klein, T. M., E. D. Wolf, R. Wu, and J. C. Sanford. 1987. High velocity microprojectiles for delivering nucleic acids into living cells. *Nature* 327:70–73.

Kleinhofs, A. 1975. DNA-hybridization studies of the fate of bacterial DNA in plants. In L. Ledoux, ed. *Genetic Manipulation with Plant Material*, pp. 461–477. New York: Plenum Press.

Kleinhofs, A., and R. M. Behki. 1977. Prospects for plant genome modification by nonconventional methods. *Ann. Rev. Genet.* 11:79–101.

Kleinhofs, A., F. C. Eden, M.-D. Chilton, and A. J. Bendich. 1975. On the question of the integration of exogenous bacterial DNA into plant DNA. *Proc. Natl. Acad. Sci. USA* 72:2748–2752.

Koukolikova-Nicola, Z., R. D. Shillito, B. Hohn, K. Wang, M. Van Montagu, and P. Zambryski. 1985. Involvement of circular intermediates in the transfer of T-DNA from *Agrobacterium tumefaciens* to plant cells. *Nature* 313:191–196.

Krens, F. A., R. M. W. Mans, T. M. S. van Slogteren, J. H. C. Hoge,, G. J. Wullems, and R. A. Schilperoort. 1985. Structure and expression of DNA transferred to tobacco via transformation of protoplasts with Ti plasmid DNA: Co-transfer of T-DNA and non T-DNA sequences. *Plant Mol. Biol.* 5:223–234.

Krens, F. A., L. Molendijk, G. J. Wullems, and R. A. Schilperoort. 1982. In vitro transformation of plant protoplasts with Ti plasmid DNA. *Nature* 296:72–74.

Langridge, J., and R. D. Brock. 1961. A thiamine-requiring mutant of the tomato. *Australian J. Biol. Sci.* 14:66–69.

Lappe, F. M., J. Collins, and C. Fowler. 1979. *Food First: Beyond the Myth of Scarcity.* New York: Ballentine.

Ledoux, L., and R. Huart. 1968. Integration and replication of DNA of *M. lysodeikticus* in DNA of germinating barley. *Nature* 218:1256–1259.

———. 1969. Fate of exogenous bacterial deoxyribonucleic acids in barley seedlings. *J. Mol. Biol.* 43:243–262.

Ledoux, L., R. Huart, and M. Jacobs. 1971. Fate of exogenous DNA in *Arabidopsis thaliana*. *Eur. J. Biochem.* 23:96–108.

———. 1974. DNA-mediated genetic correction of thiamineless *Arabidopsis thaliana*. *Nature* 249:17–21.

Ledoux, L., R. Huart, A. Ryngaert-Adriaensens, R. F. Matagne, J. P. Schlösser, and M. Jacobs. 1979. DNA-mediated correction of *Arabidopsis* thiamine mutants. *Can. J. Genet. Cytol.* 21:515–523.

Leemans, J., R. Deblaere, L. Willmitzer, H. De Greve, J. P. Hernalsteens, M. Van Montagu, and J. Schell. 1982. Genetic identification of functions of TL-DNA transcripts in octopine crown galls. *EMBO J.* 1:147–152.

Leffel, S. M., S. A. Mabon, and C. N. Stewart. 1997. Applications of green fluorescent protein in plants. *BioTechniques* 23:912–918.

Liebke, B., and D. Hess. 1977. Uptake of bacterial DNA into isolated mesophyll protoplasts of *Petunia hybrida*. *Biochem. Physiol. Pflanzen* 171:493–501.

Lippincott, J. A., and B. B. Lippincott. 1975. The genus *Agrobacterium* and plant tumorigenesis. *Ann. Rev. Microbiol.* 29:377–405.

Livi-Bacci, M. 1992. *A Concise History of World Population.* Cambridge: Balckwell.

Lurquin, P. F. 1976. Integration of exogenous DNA in plants: A hypothesis awaiting clear-cut demonstration. In D. Dudits, G. L. Farkas, and P. Maliga, eds. *Cell Genetics in Higher Plants*, pp. 77–89. Budapest: Akademiai Kiado.

————. 1977. Integration versus degradation of exogenous DNA in plants: An open question. In W. E. Cohn, ed. *Progress in Nucleic Acid Research and Molecular Biology*, pp. 161–207. New York: Academic Press.

————. 1979. Entrapment of plasmid DNA by liposomes and their interactions with plant protoplasts. *Nucl. Acids Res.* 6:3773–3784.

————. 1987. Foreign gene expression in plant cells. *Prog. Nucl. Acid Res. Mol. Biol.* 34:143–188.

————. 1997. Gene transfer by electroporation. *Mol. Biotech.* 7:5–35.

Lurquin, P. F., and R. M. Behki. 1975. Uptake of bacterial DNA by *Chlamydomonas reinhardi. Mutat. Res.* 29:35–51.

Lurquin, P. F., G. Chen, and A. J. Conner. In press. Energy dissipation is the most important parameter in plant protoplast electroporation. In Y. H. Hui, G. Khatchatourians, D. Lydiate, A. McHughen, W. K. Nip, and R. Scorza, eds. *Handbook of Transgenic Food Plants*. New York: Marcel Dekker.

Lurquin, P. F., E. Filippone, and G. Colucci. In press. Transgenic cowpea, lentil and chickpea with reporter and agronomically relevant genes. In Y. H. Hui, G. Khatchatourians, D. Lydiate, A. McHughen, K. W. Nip, and R. Scorza, eds. *Handbook of Transgenic Food Plants*, New York: Marcel Dekker.

Lurquin, P. F., and Y. Hotta. 1975. Reutilization of bacterial DNA by *Arabidopsis thaliana* cells in tissue culture. *Plant Sci. Lett.* 5:103–112.

Lurquin, P. F., and C. I. Kado. 1977. *Escherichia coli* plasmid pBR313 insertion into plant protoplasts and into their nuclei. *Mol. Gen. Genet.* 154:113–121.

Lurquin, P. F., and A. Kleinhofs. 1982. Effects on chloramphenicol on plant cells: Potential as a selectable marker for transformation studies. *Biochem. Biophys. Res. Comm.* 107:286–293.

Lurquin, P. F., and L. Márton. 1980. DNA transfer experiments with plant protoplasts and bacterial plasmids. In L. Ferenczy and G. L. Farkas, eds. *Advances in Protoplast Research*, pp. 389–405. New York: Pergamon Press.

Lurquin, P., M. Mergeay, and J. van der Parren. 1972. Banding of depolymerized DNAs in CsCl density gradients studied by computer-aided simulations. In L. Ledoux, ed. *Uptake of Informative Molecules by Living Cells*, pp. 47–50. Amsterdam: North Holland.

Lurquin, P. F., and C. Paszty. 1988. Electroporation of tobacco protoplasts with homologous and nonhomologous transformation vectors. *J. Plant Physiol.* 133:332–335.

Lurquin, P. F., and F. Rollo. 1983. Intracellular distribution of donor DNA following fusion between plant protoplasts and DNA-loaded liposomes. *Biol. Cell* 47:117–120.

Lurquin, P. F., and R. E. Sheehy. 1982. Binding of large liposomes to plant protoplasts and delivery of encapsulated DNA. *Plant Sci. Lett.* 25:133–146.

Mann, C. C. 1999a. Crop scientists seek a new revolution. *Science* 283:310–314.

———. 1999b. Genetic engineers aim to soup up crop photosynthesis. *Science* 283:314–316.

Mansour, S. L., K. R. Thomas, and M. R. Capecchi. 1988. Disruption of the proto-oncogene *int-2* in mouse embryo-derived stem cells: A general strategy for targeting mutations to non-selectable genes. *Nature* 336:348–352.

Márton, L., G. J. Wullems, P. F. Lurquin, L. Molendik, and R. A. Schilperoort. 1979. Crown gall transformation of tobacco protoplasts by Ti plasmid DNA of *Agrobacterium tumefaciens*. *Abstracts of the Fifth International Protoplast Symposium*, p. 136. Szeged, Hungary.

Márton, L., G. J. Wullems, L. Molendijk, and R. A. Schilperoort. 1979. In vitro transformation of cultured cells from *Nicotiana tabacum* by *Agrobacterium tumefaciens*. *Nature* 277:129–131.

Matthews, B. F., and D. E. Cress. 1981. Liposome-mediated DNA delivery of DNA to carrot protoplasts. *Planta* 153:90–94.

Matthysse, A. G., and A. J. Stump. 1976. The presence of *Agrobacterium tumefaciens* plasmid DNA in crown gall tumor cells. *J. Gen. Microbiol.* 95:9–16.

Matzke, M. A., and A. J. M. Matzke. 1995. Homology-dependent gene silencing in transgenic plants: How and why do plants inactivate homologous (trans)genes? *Plant Physiol.* 107:679–685.

Mazur, B., E. Krebbers, and S. Tingey. 1999. Gene discovery and product development for grain quality traits. *Science* 285:372–375.

McCabe, D. E., W. F. Swain, B. J. Martinell, and P. Christou. 1988. Stable transformation of soybean (*Glycine max*) by particle acceleration. *Bio/Technology* 6:923–926.

Memelink, J., G. J. Wullems, and R. A. Schilperoort. 1983. Nopaline T-DNA is maintained during regeneration and generative propagation of transformed tobacco plants. *Mol. Gen. Genet.* 190:516–522.

Merlo, D. J., and E. W. Nester. 1977. Plasmids in avirulent strains of *Agrobacterium*. *J. Bacteriol.* 129:76–80.

Merril, C. R., M. R. Geier, and J. C. Pettriciani. 1971. Bacterial virus gene expression in human cells. *Nature* 233:398–400.

Meyer, P., and H. Saedler. 1996. Homology-dependent gene silencing in plants. *Ann. Rev. Plant Physiol. Plant Mol. Biol.* 47:23–48.

Moffat, A. S. 1999. Crop engineering goes South. *Science* 285:370–371.

Müller, A., T. Manzara, and P. F. Lurquin. 1984. Crown gall transformation of tobacco callus cells by cocultivation with *Agrobacterium tumefaciens*. *Biochem. Biophys. Res. Comm.* 123:458–462.

Müller-Hill, B. 1996. *The lac Operon: A Short History of a Genetic Paradigm*. New York: Walter de Gruyter.

Neumann, E., M. Schaefer-Ridder, Y. Wang, and P. H. Hofschneider. 1982. Gene transfer into mouse lyoma cells by electroporation in high electric fields. *EMBO J.* 1:841–845.

Nussaume, L., M. Vincentz, and M. Caboche. 1991. Constitutive nitrate reductase: A dominant conditional marker for plant genetics. *Plant J.* 1:267–274.

Odell, J., P. Caimi, B. Sauer, and S. Russell. 1990. Site-directed recombination in the genome of transgenic tobacco. *Mol. Gen. Genet.* 223:369–378.

Ohyama, K., O. L. Gamborg, and R. A. Miller. 1972. Uptake of exogenous DNA by plant protoplasts. *Can. J. Bot.* 50:2077–2080.

Ooms, G., P. J. Hooykaas, G. Moleman, and R. A. Schilperoort. 1981. Crown gall plant tumors of abnormal morphology, induced by *Agrobacterium tumefaciens* carrying mutated octopine Ti plasmids: Analysis of T-DNA functions. *Gene* 14:33–50.

Otten, L., H. De Greve, J. P. Hernalsteens, M. Van Montagu, O. Schieder, J. Straub, and J. Schell. 1981. Mendelian transmission of genes introduced into plants by the Ti plasmids of *Agrobacterium tumefaciens*. *Mol. Gen. Genet.* 183:209–213.

Otten, L. A. B. M., and R. A. Schilperoort. 1978. A rapid micro scale method for the detection of lysopine and nopaline dehydrogenase activities. *Biochem. Biophys. Acta* 527:497–500.

Owens, L. D. 1979. Binding of ColE1-kan plasmid DNA by tobacco protoplasts. *Plant Physiol.* 63:683–686.

Paszkowski, J., M. Baur, A. Bogucki, and I. Potrykus. 1988. Gene targeting in plants. *EMBO J.* 7:4021–4026.

Paszkowski, J., R. D. Shillito, M. Saul, V. Mandak, T. Hohn, B. Hohn, and I. Potrykus. 1984. Direct gene transfer to plants. *EMBO J.* 3:2717–2722.

Perkins, E. J., M. P. Gordon, O. Caceres, and P. F. Lurquin. 1990. Organization and sequence analysis of the 2,4-dichlorophenol hydroxylase and dichlorocatechol oxidative operons of plasmid pJP4. *J. Bacteriol.* 172:2351–2359.

Perkins, E. J., and P. F. Lurquin. 1988. Duplication of 2,4-dichlorophenoxyacetic acid monooxygenase gene in *Alcaligenes eutrophus* JMP134(pJP4). *J. Bacteriol.* 170:5669–5672.

Pfeiffer, P., and T. Hohn. 1983. Involvement of reverse transcription in the replication of cauliflower mosaic virus: A detailed model and test of some aspects. *Cell* 33:781–789.

Potrykus, I., J. Paszkowski, M. W. Saul, J. Petruska, and R. D. Shillito. 1985. Molecular and general genetics of a hybrid foreign gene introduced into tobacco by direct gene transfer. *Mol. Gen. Genet.* 199:169–177.

Rédei, G. P., G. Acedo, H. Weingarten, and L. D. Kier. 1976. Has DNA corrected genetically thiamineless mutants of *Arabidopsis*? In D. Dudits, G. L. Farkas, and P. Maliga, eds. *Cell Genetics in Higher Plants*. Budapest: Akadémiai Kiadó.

Rochaix, J.-D., and J. van Dillewijn. 1982. Transformation of the green alga *Chlamydomonas reinhardii* with yeast DNA. *Nature* 296:70–71.

Rollo, F., M. G. Galli, and B. Parisi. 1981. Liposome-mediated transfer of DNA to carrot protoplasts: A biochemical and autoradiographic analysis. *Plant Sci. Lett.* 20:347–354.

Rollo, F., and R. Hull. 1982. Liposome-mediated infection of turnip protoplasts with turnip rosette virus and RNA. *J. Gen. Virology* 359–363.

Rouze, P., A. Deshayes, and M. Caboche. 1983. Use of liposomes for the transfer of nucleic acids: Optimization of the method for tobacco mesophyll protoplasts with tobacco mosaic virus RNA. *Plant Sci. Lett.* 31:55–64.

Sabri, N., B. Pelissier, and J. Teissie. 1996. Transient and stable electrotransformations of intact black Mexican sweet maize cells are obtained after preplasmolysis. *Plant Cell Rep.* 15:924–928.

Sander, E. 1964. Evidence of the synthesis of a DNA phage in leaves of tobacco plants. *Virology* 24:545–551.

———. 1967. Alteration of fd phage in tobacco leaves. *Virology* 35:121–130.

Schell, J. 1975. The role of plasmids in crown gall formation by *A. tumefaciens.* In L. Ledoux, ed. *Genetic Manipulation with Plant Material*, pp. 163–181. New York: Plenum Press.

Serageldin, I. 1999. Biotechnology and food security in the 21st century. *Science* 285:387–389.

Sheehy, R. E., M. Kramer, and W. R. Hiatt. 1988. Reduction of polygalacturonase activity in tomato fruit by antisense RNA. *Proc. Natl. Acad. Sci. USA* 85:8805–8809.

Sheehy, R. E., and P. F. Lurquin. 1983. Targeting of large liposomes with lectins increases their binding to plant protoplasts. *Plant Physiol.* 72:386–390.

Shillito, R. D., M. W. Saul, J. Paszkowski, M. Muller, and I. Potrykus. 1985. High efficiency direct gene transfer to plants. *Bio/Technology* 3:1099–1104.

Smith, H. O., and D. B. Danner. 1981. Genetic transformation. *Ann. Rev. Biochem.* 50:41–68.

Smith, H., R. A. McKee, T. H. Attridge, and D. Grierson. 1975. Studies on the use of transducing bacteriophages as vectors for the transfer of foreign genes to higher plants. In L. Ledoux, ed. *Genetic Manipulations with Plant Materials*, pp. 551–563, New York: Plenum Press.

Somerville, C., and S. Somerville. 1999. Plant functional genomics. *Science* 285:380–383.

Songstad, D. D., D. A. Somers, and R. J. Griesbach. 1995. Advances in alternative DNA delivery techniques. *Plant Cell, Tissue and Organ Culture* 40:1–15.

Soyfer, V. N., and Y. B. Titov. 1981. The absence of a definite gene-specific effect in wheat after seed treatment with exogenous DNA. *Mol. Gen. Genet.* 182:361–363.

Stachel, S. E., E. Messens, M. Van Montagu, and P. Zambryski. 1985. Identification of the signal molecules produced by wounded plant cells that activate T-DNA transfer in *Agrobacterium tumefaciens. Nature* 318:624–628.

Stachel, S. E., and E. W. Nester. 1986. The genetic and transcriptional organization of the vir region of the A 6 Ti plasmid of *Agrobacterium tumefaciens. EMBO J.* 5: 1445–1454.

Stachel, S. E., B. Timmerman, and P. Zambryski. 1986. Generation of single-stranded T-DNA molecules during the initial stages of T-DNA transfer from *Agrobacterium tumefaciens* to plant cells. *Nature* 322:706–712.

Stachel, S. E., and P. C. Zambryski. 1986. VirA and virG control the plant-induced activation of the T-DNA transfer process of *A. tumefaciens. Cell* 46:325–333.

Stent, G. S. 1972. Prematurity and uniqueness in scientific discovery. *Scientific American* 227:84–93.

Stougaard, J. 1993. Substrate-dependent negative selection in plants using a bacterial cytosine deaminase gene. *Plant J.* 3:755–761.

Streber, W. R., and L. Willmitzer. 1989. Transgenic tobacco plants expressing a bacterial detoxifying enzyme are resistant to 2,4-D. *Bio/Technology* 7:811–816.

Stroun, M., P. Anker, P. Charles, and L. Ledoux. 1967. Translocation of DNA of bacterial origin in *Lycopersicum esculentum* by ultracentrifugation in caesium chloride gradient. *Nature* 215:975–976.

Stroun, M., P. Anker, and L. Ledoux. 1967. DNA replication in *Solanum lycopersicum esc.* after absorption of bacterial DNA. *Curr. Mod. Biol.* 1:231–234.

Sudhakar, D., L. T. Duc, B. B. Bong, P. Tinjuanjung, S. B. Maqbool, M. Valdez, R. Jefferson, and P. Christou. 1998. An efficient rice transformation system utilizing mature seed-derived explants and a portable, inexpensive particle bombardment device. *Transgenic Res.* 7:289–294.

Suzuki, M., and I. Takebe. 1976. Uptake of single-stranded bacteriophage DNA by isolated tobacco protoplasts. *Z. Pflanzenphysiol.* 78:421–433.

———. 1978. Uptake of double-stranded bacteriophage DNA by isolated tobacco leaf protoplasts. *Z. Pflanzenphysiol.* 89:297–311.

Svab, Z., and P. Maliga. 1993. High-frequency plastid transformation in tobacco by selection for a chimeric *aadA* gene. *Proc. Natl. Acad. Sci. USA* 90:913–917.

Thomashow, M. F., R. Nutter, K. Postle, M.-D. Chilton, F. R. Blattner, A. Powell, M. P. Gordon, and E. W. Nester. 1981. Recombination between higher plant DNA and the Ti plasmid of *Agrobacterium tumefaciens. Proc. Natl. Acad. Sci. USA* 77:6448–6452.

Thompson, C. J., N. R. Movva, R. Tizard, R. Crameri, J. E. Davies, M. Lauwereys, and J. Botterman. 1987. Characterization of the herbicide-resistance gene *bar* from *Streptomyces hygroscopicus. EMBO J.* 6:2519–2523.

Timmerman, B., M. Van Montagu, and P. Zambryski. 1988. Vir-induced recombination in *Agrobacterium*. Physical characterization of precise and imprecise T-circle formation. *J. Mol. Biol.* 203:373–384.

Turbin, N. V., V. N. Soyfer, N. A. Kartel, N. M. Chekalin, Y. L. Dorohov, Y. B. Titov, and K. K. Cieminis. 1975. Genetic modification of the *waxy* character in barley under the action of exogenous DNA of the wild variety. *Mutat. Res.* 27:59–68.

Uchimiya, H., and T. Murashige. 1977. Quantitative analysis of the fate of exogenous DNA in *Nicotiana* protoplasts. *Plant Physiol.* 59:301–308.

Uchimiya, J., and H. Harada. 1981. Transfer of liposome-sequestering plasmid DNA into *Daucus carota. Plant Physiol.* 68:1027–1030.

Ursic, D., J. D. Kemp, and J. P. Helgeson. 1981. A new antibiotic with known resistance factors, G418, inhibits plant cells. *Biochem. Biophys. Res. Comm.* 101:1031–1036.

van der Krol, A. R., P. E. Lenting, J. Veenstra, I. M. van der Meer, R. E. Koes, A. G. M. Gerats, J. N. M. Mol, and A. R. Stuitje. 1988. An antisense chalcone synthase gene in transgenic plants inhibits flower pigmentation. *Nature* 333:866–869.

Van Larebeke, N., G. Engler, M. Holsters, I. Zaenen, R. A. Schilperoort, and J. Schell. 1974. Large plasmid in *Agrobacterium tumefaciens* essential for crown gall-inducing ability. *Nature* 252:197–170.

Van Larebeke, N., C. Genetello, J. Schell, R. A. Schilperoort, A. K. Hermans, J. P. Hernalsteens, and M. Van Montagu. 1975. Acquisition of tumor-inducing ability by non-oncogenic agrobacteria as a result of plasmid transfer. *Nature* 255:742–743.

Van Slogteren, G. M. S., P. J. J. Hooykaas, and R. A. Schilperoort. 1984. Expression of Ti-plasmid genes in monocotyledonous plants infected with *Agrobacterium tumefaciens*. *Nature* 311:763–764.

Van Vliet, F., M. De Beuckeleer, A. Depicker, G. Engler, M. Lemmers, M. Holsters, M. Van Montagu, and J. Schell. 1980. The *Agrobacterium* Ti plasmid as a host vector system for introducing foreign DNA in plant cells. *Nature* 287:654–656.

Wallroth, M., A. G. M. Gerats, S. G. Rogers, R. T. Fraley, and R. B. Horsch. 1986. Chromosomal localization of foreign genes in *Petunia hybrida*. *Mol. Gen. Genet.* 202:6–15.

Ward, E. R., and W. M. Barnes. 1988. Vir D2 protein of *Agrobacterium tumefaciens* very tightly linked to the 5′ end of T-strand DNA. *Science* 242:927–930.

Watson, B., T. C. Currier, M. P. Gordon, M.-D. Chilton, and E. W. Nester. 1975. Plasmid required for virulence of *Agrobacterium tumefaciens*. *J. Bacteriol.* 123:255–264.

Weaver, R. F., and P. W. Hedrick. 1991. *Basic Genetics*. Oxford: Wm. C. Brown.

Weber, G., S. Monajembashi, K. O. Greulich, and J. Wolfrum. 1988. Injection of DNA into plant cells with a UV laser microbeam. *Naturwissenschaften* 75:35–36.

Wendt-Gallitelli, M. F., and I. Dobrigkeit. 1973. Investigations implying the invalidity of octopine as a marker for transformation by *Agrobacterium tumefaciens*. *Z. Naturforschung.* 28:768–771.

Wigler, M., R. Sweet, G. K. Sim, B. Wold, A. Pellicer, E. Lacy, T. Maniatis, S. Silverstein, and R. Axel. 1979. Transformation of mammalian cells with genes from procaryotes and eucaryotes. *Cell* 16:777–785.

Willmitzer, L., M. De Beuckeleer, M. Lemmers, M. Van Montagu, and J. Schell. 1980. DNA from Ti plasmid present in nucleus and absent from plastids of crown gall plant cells. *Nature* 287:359–361.

Wolffe, A. P., and M. A. Matzke. 1999. Epigenetics: Regulation through repression. *Science* 286:481–486.

Xonocostle-Cázares, B., Y. Xiang, R. Ruiz-Medrano, H.-L. Wang, J. Monzer, B.-C. Yoo, K. C. McFarland, V. R. Franceschi, and W. J. Lucas. 1999. Plant paralog to viral movement protein that potentiates transport of mRNA into the phloem. *Science* 283:94–98.

Yadav, N. S., K. Postle, R. K. Saiki, M. F. Thomashow, and M.-D. Chilton. 1980. T-DNA of crown gall teratoma is covalently joined to host plant DNA. *Nature* 287:458–461.

Yadav, N. S., J. Vanderleyden, D. R. Bennett, W. M. Barnes, and M.-D. Chilton. 1982. Short direct repeats flank the T-DNA on a nopaline Ti plasmid. *Proc. Natl. Acad. Sci. USA* 79:6322–6326.

Yamaoka, N., I. Furusawa, and M. Yamamoto. 1982. Infection of turnip protoplasts with cauliflower mosaic virus DNA. *Virology* 122:503–505.

Yang, F., D. J. Merlo, M. P. Gordon, and E. W. Nester. 1980. Plasmid DNA of *Agrobacterium tumefaciens* detected in a presumed habituated tobacco cell line. *Molec. Gen. Genet.* 179:223–226.

Yang, F., A. L. Montoya, D. J. Merlo, M. H. Drummond, M.-D. Chilton, E. W. Nester, and M. P. Gordon. 1980. Foreign DNA sequences in crown gall teratomas and their fate during the loss of the tumorous traits. *Mol. Gen. Genet.* 177:707–714.

Zaenen, I., N. Van Larebeke, H. Teuchy, M. Van Montagu, and J. Schell. 1974. Supercoiled circular DNA in crown gall inducing *Agrobacterium* strains. *J. Mol. Biol.* 86:109–27.

Zambryski, P., H. Joos, C. Genetello, J. Leemans, M. Van Montagu, and J. Schell. 1983. Ti plasmid vectors for the introduction of DNA into plant cells without alteration of their normal regeneration capacity. *EMBO J.* 2:2143–2150.

Index